URBAN AGRICULTURE

URBAN AGRICULTURE

– Author –
Gurdeep Singh
Mahi Kohli
Jaspal Singh

ISBN: 978-81-19283-06-4 (Hardback)

Published by : **WORDPEN ACADEMICS**
4736/23, Ansari Road, Darya Ganj,
New Delhi – 110 002
Phone: +91-011-35004336
E-mail: info@wordpenacademics.com

Printed at **:** **Replika Press Pvt. Ltd.**

Exclusive Distributor: ***Bio-Green Books, New Delhi***

Preface

Urban agriculture refers to the practice of growing food, raising livestock, or producing other agricultural products in urban areas. With urbanization and population growth, urban agriculture has emerged as a sustainable solution to meet the increasing demand for food, reduce food insecurity, and promote environmental sustainability in cities. Urban agriculture has gained popularity in recent years due to the growing interest in local food production, food security, and the environmental impact of food production and distribution. By producing food locally, urban agriculture can reduce the carbon footprint associated with long-distance transportation and distribution of food. Moreover, urban agriculture can promote local economic development by creating jobs and supporting small businesses. Urban agriculture comes in many forms, from community gardens and rooftop farms to hydroponic and aquaponic systems.

Urban agriculture also has the potential to improve the environment and enhance the livability of urban areas. Green roofs and rooftop gardens can reduce the urban heat island effect by providing shade and absorbing heat, while also improving air quality by capturing pollutants. Moreover, urban agriculture can promote biodiversity in urban areas by providing habitat for insects and wildlife. Despite its potential benefits, urban agriculture faces many challenges, including limited access to land, inadequate infrastructure, and regulatory barriers. For example, zoning regulations and building codes may restrict the use of rooftops and vacant lots for agriculture, and access to water and other resources can be limited in urban areas. To overcome these challenges, urban agriculture advocates and policymakers are working to promote policies and initiatives that support urban agriculture, such as community land trusts, zoning ordinances, and incentives for urban agriculture businesses. Moreover, public-private partnerships can help to provide funding, technical assistance, and other resources to support urban agriculture. Urban agriculture offers a sustainable and innovative solution to meet the challenges of food insecurity, environmental degradation, and urbanization. By promoting local food production, reducing food miles, and improving the livability of urban areas, urban agriculture has the potential to transform our cities into more sustainable and resilient communities.

It covers various aspects of urban agriculture, including case studies, sustainability, and environmental impact and also provides an overview of urban agriculture, its history, and its importance in modern times. The book emphasizes the importance of sustainable urban agriculture practices, including the use of organic and locally sourced materials, minimizing waste, and conserving resources like water and energy. It examines the environmental impact of urban agriculture, including its potential to reduce carbon emissions, improve air and water quality, and increase biodiversity in urban areas. The book also explores the relationship between urban agriculture and land use, particularly in small islands and peri-urban areas and also examines the relationship between population growth and agricultural land. This book is highly useful to students, teachers, researchers, policymakers, environmentalists, community activists, and educators interested in sustainable urban development and agriculture.

We are grateful to all those persons as well as various books, manuals, periodicals, magazines, journals etc. that helped in the preparation of this book. In spite of the best efforts, it is possible that some errors may have occurred into the compilation and editing of the book. Further queries, constructive suggestions and criticisms for the improvement of the book are always welcomed and shall be thankfully acknowledged.

Gurdeep Singh
Mahi Kohli
Jaspal Singh

Contents

Chapter 1

Introductory Chapter

Urban expansions endanger agricultural, natural, and green areas. Therefore, urban agriculture is suggested as an important solution toward minimizing urban pollution and allowing more beautiful scenery in our cities. Additionally, urban agriculture could be an important source of income. The yielded agricultural products can be sold at farmer's markets in urban locations. Urban agriculture is practiced in a variety of forms and scales. Practices range from the production of crops in small to large lots, vertical production on walls, windows (window farms), and rooftops (green roofs). Therefore, this book was composed to provide useful information about urban agriculture which includes urban farming, urban gardens, domestic gardens, urban agriculture, intra-urban agriculture, peri-urban agriculture, peri-urban farming, green roofs, window farm, irrigated urban agriculture, urban landscaping, landscape, suburban livestock farming, urban beekeeping, food security, urbanization, land-use change, agricultural systems, and land-use planning.

This book introduces several concepts of urban agriculture with case study examples representing practices. These case studies highlight farmer's markets, urban gardens, walls, rooftops, and roof deck where crop varieties are grown on top of buildings. The diversity of these activities exhibits transferable concepts to many locations around the world. The challenges for these practices include adaptability of crops, structural and community support, and the presence of a viable market.

In order to evaluate the agricultural systems in the context of urban agriculture, economic models are highly required. Accordingly, economic models of urban gardening are presented in this book, such as the avoided costs model (ACM) and the business model (BM). The main economic differences between the economic models are exemplified. An analysis of the phenomenon of urban gardening as a business model for small family homes or allotment gardens is presented in this book.

On the other hand, important methodologies to investigate the evolution of land uses are presented in this book, where cutting-edge technologies are implemented, such as the geographic information systems (GIS) and the orthophotomaps to evaluate the spatial planning of the land uses. Subsequently, this provides important understanding of the impacts of land uses on ecosystems in the frame of sustainable development. As a result, sustainable spatial development scenarios are proposed.

In this book, a descriptive approach for the functionality of peri-urban agricultural systems is presented, where this approach focuses on multifunctionality and the goods and services of agricultural systems. This book shows a wide variety of functions that can be grouped according to their economic dimension, social dimension, and environmental dimension. A methodology to quantify the functionality of peri-urban agricultural systems by means of indicators is proposed in this book.

This book provides a methodology to investigate the effect of population in determining which areas see population increases and are under pressure. The methodology is to determine population in the neighborhoods, the distribution of rural areas in the city plans in the neighborhoods, and for this reason implements the geographical information systems to conduct the analyses. This methodology investigates further the existence of a negative or positive interaction between population and agricultural areas. This methodology leads to describe the problems of rural areas located in urban areas and indicate their status according to the population.

Further important agricultural systems are implemented in urban areas which are the aquaculture systems. Therefore, this hot topic is discussed in this book. The modern aquaculture technologies are made in recirculating systems, which require the use of high-performance methods for recirculating the treated wastewater. This book presents the wastewater treatment processes in intensive aquaculture systems.

Chapter 2

Urban Agriculture Case Studies in Central Texas: From the Ground to the Rooftop

Abstract

Urban agriculture is practiced in various forms and scales. Practices range from the production of edibles in small to large lots or plazas to vertical production on walls and rooftops. Produce is grown in rural locations and sold at farmers markets in urban locations or grown onsite. Efforts to produce, maintain and sell products of urban agriculture involve many participants and leaders from multiple disciplines. This chapter highlights an introduction to several urban agriculture concepts and case study examples representing activities in Central Texas, home of the Texas A&M University. These case studies highlight a modular pavilion type farmers market and urban garden in downtown Bryan, Texas, designed and built by students and faculty collaborations. The farmers market is designed to be a flexible structure to accommodate current and future needs. A rooftop crop pilot study at Texas A&M on walls and roof deck highlights the varieties grown on top of a four-storey building. The diversity of activities taking place in Central Texas exhibits concepts transferable to many locations across the world. The challenges for these projects include adaptability of crops to the Central Texas climate, structural and community support and the presence of a viable market for locally grown produce.

Keywords: Architecture, Landscape Architecture, Design/Build, Interdisciplinary Learning, Rooftop Farming

1. Introduction

Urban agriculture exists in a variety of forms worldwide. There are various ways how the food is grown and sold. Production, for example, varies from large-scale plots on private or corporate property to small-scaled applications for private or public consumption. Crops are grown in urban areas for sales in local farmers markets, for private use, for restaurants or grown through community- or municipal-supported efforts. Distribution varies from private sales from the back of a truck roadside to public selling of food, crops and marketable goods in farmers markets or local stores. Urban agriculture exists on private and public land as well as on rooftops exposed to the elements or concealed in greenhouses. Urban agriculture exists as permanent ventures as well as temporary or seasonal events by small groups, organizations or private individuals. Urban agriculture has historically had strong ties to small-scale grassroots movements and ties to agrarian beginnings, but recently in North America and abroad, there is greater need for people living in cities to reconnect with nature, taste fresh seasonal produce, socialize and learn [1].

One unique example of a municipal level (top down) production of crops outside North America is the Edible City Project at the village of Andernach, Germany [2]. There, the public right-of-way is used in the town Center to grow fruits, vegetables, cut flowers and some produce such as fresh chicken eggs. Permaculture concepts are implemented to maximize healthy yields and reduce environmental risks. The yields are free to the public via self-picking, but programmed harvests are organized by local unemployed workers under the direction of a city program. There, the workers earn a minimum wage and learn marketable skills such as growing food, harvesting, selling and managing operations. The food is produced organically at a small scale in the city proper, but a larger permaculture farm exists outside the village to sustain greater quantities of produce for the local outlet. Proceeds go to reinvesting in the program. Seasonal crops and cut flowers are sold in the downtown market (**Figure 1**).

Figure 1. Local goods grown in the public right-of-way in Andernach, Germany, for sale in a downtown market (photo by B. Dvorak).

Across the United States, there are many examples of a variety of forms of urban agriculture including farmers markets, selling of local produce in commercial stores, co-ops and along roadsides in small-scale structures or the back of a truck. Urban agriculture in Texas exists in similar forms and is experiencing a rebirth. Texas' agricultural roots extend back to their formative years and are well known for its cattle and cotton. Urban centers were traditionally the life of this cultural exchange in the Public Square or market. Today, food in the United States is not centrally distributed in a city Center but is sold widely across supermarkets dispersed widespread across metro areas. The popularity and revival of famers markets suggest that citizens enjoy getting outside, visiting with local producers and learning about the current farming practices and supporting local producers.

The agricultural roots of Central Texas are found in the heart of the Brazos Valley. Brazos County, Texas, includes a population of over 200,000 and is named for the Brazos River, which forms its western border and forms the Brazos Valley. The significance of the river to Spanish explorers can be seen in its name *el Rio de los Brazos de Dios,* which translates to "the River of the Arms of God," which influenced the naming of the county to Brazos County. The river contains a very rich soil that it came just second place to soil from the Nile river in Egypt in an international competition about soil fertility [3]. The Brazos County was formed in 1841 and organized in 1843. Originally one of the state's poorer counties, in 1870 it donated 2416 acres of land to create Texas A&M University, which has since enabled the county to be among the state's most financially stable [4].

The authors, also faculty members at Texas A&M University's College of Architecture, have lead and directed numerous interdisciplinary learning initiatives that focused on urban agriculture. These initiatives investigated the role of both architecture and landscape architecture disciplines in activating real-life experimental projects that involved graduate and undergraduate students and collaborated with community members and local farmers. In this chapter, case studies are presented and discussed: a farmers market pavilion structure in the City of Bryan, Texas, and a rooftop and wall garden pilot project on campus.

The Central Texas case studies are designed and build — or currently under development — by students and faculty from Texas A&M University and in partnerships with real clients and government officials. Our inquiry stems from an interdisciplinary approach to collaborate, educate and disseminate knowledge to both college level students and our community citizens at large about urban agriculture and farming. The systematic approach of research and applications is correlated with the current industry trend of Integrated Project Delivery method (IPD) where group decision-making from both industry experts and design educators takes place in the early phases of the design process. Our research is experimental in its nature and requires testing of ideas through the physical realization of the proposed community projects.

2. Farmers markets

The U.S. Department of Agriculture estimates there are over 8100 farmers markets nationwide, a jump of almost 5000 from the previous decade [5]. The recent success in creating new businesses, increasing vendors and public partnerships was achieved in two recently built farmers markets in North America, one in Covington, Virginia, and the other in Bertie County, North Carolina. Both of these projects were surprisingly designed and built by students and architecture faculty members with community participation, and established the role of service learning initiatives that involve students to contribute to the public good and community development. These two case studies inspired us to lead the efforts for a third one in Central Texas by framing the opportunity for our students and integrating a unique learning experience to their design curriculum.

In this section, three case studies of pavilion type farmers markets are presented and discussed: Covington Farmers Market in Southwest Virginia, Bertie County Farmers Market in North Carolina and Bryan Urban Farmers Market in the Brazos Valley in Texas. All case studies are designed and build — or currently under development — by students and faculty and in partnerships with real clients and government officials.

The idea for this project was to design and build a pavilion type structure that can serve users beyond the selling and buying activities of a typical farmers market and that the building itself becomes a landmark of its community, demonstrating the power of architectural design, sustainability thinking and community partnership. The land and the structure would be paired together to propose a solution for multi-use development of the dedicated properties. At farmers markets, most produce vendors use simple, generic white canopies that need to be light-weight and portable, which also means that vendors have to get creative when they anchor them to the ground (no stakes allowed). Most vendors cannot afford a custom tent, so they are all white, with flimsy signage [6]. Often as is the situation in Bryan, Texas, a parking lot serves as a host site for farmers markets. Contemporary solutions such as the Lafayette Gardens in Detroit, Michigan, and others demonstrate that the integration of structures and site can make for dynamic relationships and activate the underutilized spaces. Our goal was to design and build a permanent pavilion structure as part of the ongoing community service design/build program offering high-impact design/build initiatives. We worked with the city of Bryan to secure a site in downtown to extend the development of the urban farming model. We have established connections with key players and closely worked with them and the city of Bryan to realize the project. The inquiry for students was to design and build a farmers market for the Bryan/College Station community. Students visited the current markets found on typical Google search and revisited the question of material sourcing and harvesting and tie that to their design proposal. Two major issues with farmers markets in the BCS area included parking for visitors and timing of operation. Some of the questions that students were challenged with were as follows:

- All goods sold at this market required to be produced within a 100-mile radius; could this distance become also a goal for the procurement of construction materials?

- How does the building meet the sky and the ground? How does it meet the street and the landscape around it? Far and near?
- What role the building will play in the community when there is no buying and selling? What kind of night lighting conditions could make the building a lantern in the dark for example?

2.1. A unique community project collaboration

Since 1960, there has been no significant structure on site for a farmers market in the Brazos County, which constitutes Bryan and College Station area (**Figure 2**) [7]. In the spring of 2015, a group of second-year architecture students designed a farmers market for the BCS community. The students started by researching, learning and then evaluating the existing situation of the farmers and local markets to understand how and when vendors sell their produce. Making use of local material sourcing and harvesting was the approach integrated into their design proposal. The interdisciplinary approach of thinking was achieved by collaborating with landscape architecture, structural engineering and construction science students. The design proposal was developed, engineered, priced and prepared for construction by the end of fall 2016 and a full construction document set will be ready to build the market. Students will utilize the Architecture Fabrication Facility at their university to prefabricate the building's components, then transport and assemble them onsite. The site design and development took place during the 2015–2016 academic year.

Students had an in-depth hands-on rich learning experience, which is based on an active participation from students and peer-learning principals of funding, designing, engineering, management, fabrication, production planning and construction. The following case studies proved that the integrated experience is outstanding and students graduate with a rich understanding of interdisciplinary collaboration. Students will be able to understand the "value" that other disciplines bring to the teamwork and learn to think as collaborators. The project is an active and dynamic learning experience in planning, budgeting, scheduling, design, construction management and community engagement. A group of farmers were an integral part of that renewed discussions exploring possibilities for design and construction of a covered, multi-function pavilion in the vacant area north of downtown Bryan to be used for farmers' market, as well as other group activities. Students worked toward identifying design features that would be critical or desirable for vendors. Farmers' members participated with their input to the preliminary design process and attended design reviews with students. Community members believed that a covered and attractive pavilion for a farmers market could be a major asset in the continuing redevelopment of downtown Bryan, as well as encouraging a more robust farmers market culture in the community. The community surrounding Texas A&M campus could sustain the growth of the locally grown fresh-food movement that is second to none in Texas.

Figure 2. City of Bryan Original Farmers Market Circa 1931 (photo from Downtown Bryan Association).

2.2. Case studies: small farmers markets reshaping community development

2.2.1. Covington Farmers Market, Virginia

Covington is a small town in Southwest Virginia that shares similarities with other towns across the United States, especially with Bryan, Texas. The working-class heritage town was built on railroads and resource extraction activities surrounded by beautiful mountains as its backdrop. The farmers market was formed with no resources by a group of local farmers. They had been operating on an open parking lot for some time but lacked a shelter and the means to realize one. The small town was all of the sudden on the map of design magazines and national news when a student-led team built a small farmers market in downtown. The market has brought new life to downtown Covington, and has been featured by Architect Magazine and received awards from the Virginia Society of the American Institute of Architects and others [8].

Former students of Auburn University's Rural Studio, Keith and Marie Zawistowski, took their passion to Virginia Tech School of Architecture and Design. In their first design/build project, students were asked to design and build a farmers market in Covington, Virginia. After studying and researching existing markets through literature and site visits, each student developed an iterative for the project. After several rounds of extracting principals and converging ideas, students arrived at one scheme to build and it was through that collaborative design process that all students owned the design idea rather than the chosen scheme. The integrated nature of the design process insured a healthy development throughout the duration of the project and minimized conflicts between students [8].

The market, which opened in late May 2011, was conceived as three parts: Ground Plane, Occupied Space and Pavilion Roof (**Figure 3**). Offsite prefabrication was a key factor in dissecting the market into 10′ wide sections, which were put together at the fabrication facility of the university, then transported and assembled on site. The clarity of the project components gave much appreciation to the articulation of its architectural elements. The roof stands as the highlighted figure of the project with its sculptural form, the floor as a floating performance stage and the slender steel posts are carefully inserted in between the two planes and autonomously rising. The project was a yearlong process that focused on the research, development and implementation of innovative construction methods and architectural designs.

Figure 3. Covington Farmers Market, Virginia (used with permission).

2.2.2. Bertie County Farmers Market, North Carolina

Bertie County is a place dominated by rural poverty. It has a median household income of $31,194. Its school system is suffering from major issues: lack of interest from students, very low passing rates and young generation of low-wage workers and farmers who lost hope in education. Studio H worked with the school board members to introduce a new curriculum that empowered students by design education [9].

Figure 4. Bertie County Farmers Market, North Carolina (used with permission).

The Windsor Super Market is the only farmers market pavilion in the country designed and built entirely by high school students. The first Studio H project, by 13 high school juniors from Bertie County, North Carolina, constructed the 2000-square-foot structure. Our students spent two semesters and the following summer researching, prototyping, engineering and building the structure, as well as spearheading the launch of the local farmers market association in their hometown of 2000 people. The story of the Windsor Market is told in a documentary film, "If You Build It" [9].

The Windsor Market pavilion was featured in Architectural Record and on NPR's The Story with Dick Gordon, and has created 2 new businesses and 15 new jobs since its opening in October 2011 (**Figure 4**). Students learned a wide range of design, drafting and building skills, applying them to a series of projects for their community. The Mayor gave the Project H students the key to the city — the second key to the city ever given out. The sense of pride that grew among the students spread to their families and out to the wider community. The school system in Bertie County is suffering from major issues; lack of interest from students, very low passing rates and young generation of low wage workers and farmers who lost hope in education. Studio H worked with the school board members to introduce a new curriculum that empower students by design education.

2.2.3. *Bryan Urban Farmers Market, Texas*

Downtown Bryan is passionate about supporting and advancing commerce, culture and community. They actively work toward these goals through economic development, support of local art and culture and community engagement. This environment has created the perfect conditions for the proposed farmers market building by the College of Architecture at Texas A&M students, which will be the highlight of the initiative and will demonstrate the power of design to the larger public. The project will be the seed for future community engagements and interventions by the COA interdisciplinary design/build group. As the case with the previous case studies, jobs are expected to be generated as well as new businesses and cultural activities. The design of a covered, multi-function pavilion to be used for farmers' market, as well as other group activities, will provide permanent pavilion that is attractive for both vendors and shoppers.

Our students' design for the market was derived from the knowledge gained from the previous projects. Sourcing and tectonic details were at the forefront of conceptual design. A farmers market is a very humble entity that leaves much room for tectonic expression of simple construction methods. Essentially, the design derives from a tent and a system of modularity that simplifies the construction. The complete design of the farmer's market encompasses the entire site. The modular unit is designed for the farmer and the community. With a 15′.6″ × 16′.6″ base, the module allows for easy flow underneath and ample space to set up a booth in which to sell produce (**Figure 5**). Entirely constructed of true 2 × 6 cedar boards and custom steel plates, the module lends itself to ease of construction and deconstruction. The connections throughout the design echo the idea of creating connections within the community. Every wood member is connected to one another by a steel plate, emphasizing the actual structural connection itself well. Stretched over the top of this wood and steel frame is translucent polycarbonate layer. With a life span of over 50 years, the layer would be diffusing all direct sunlight and providing the maximum shade. With a clearance height of 10′, farmers are also able to drive their trucks within the structure to allow ease of accessibility in unloading and loading the produce.

The market is a community development project with a focus on green spaces. Incorporated into this site is our modular structure that offers a dynamic range of implementation whether it be a farmers market, concert, wedding or birthday party. Just being north of downtown Bryan, Texas, the area lends itself to both daytime and nighttime foot traffic. This will allow the market to be used almost the entire year, creating a constant income for the city. Located near shopping centers, the site will become a landmark and a central location for the public to reside in. The structure itself only covers 48% of the total square footage, allowing the other space to be used for green design such as planters or water recycling (**Figures 5** and **6**). We hope as well that the city of Bryan, Texas, is not the only community to utilize this module. Offering the kit of parts to other communities would allow a universal market structure to be built in a number of different orientations around the state.

Figure 5. Bird's eye view of the proposed modular market.

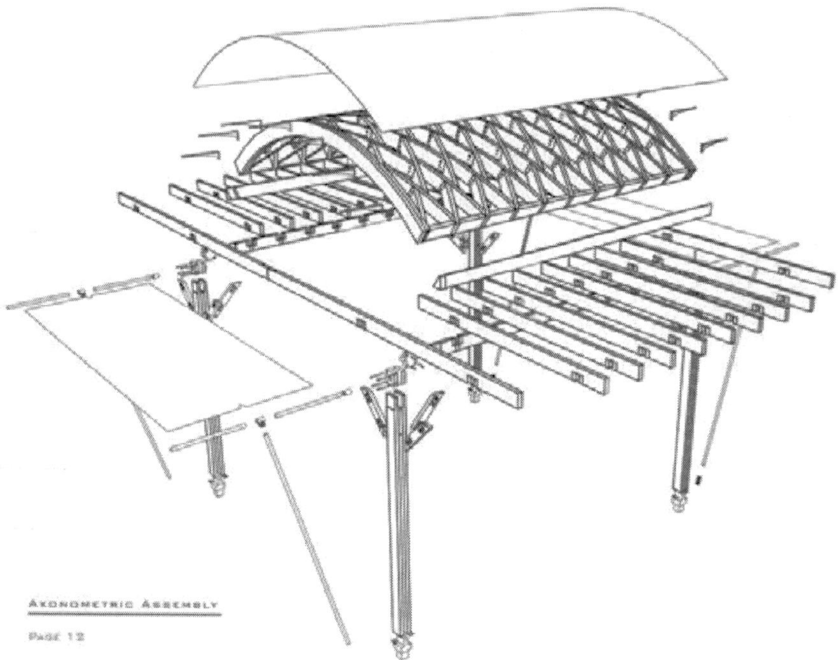

Figure 6. Assembly of the modular unit.

2.3. Conclusion

Urban farming or urban agriculture is quickly becoming an integral part of today's modern living, but what is the contribution of architecture and landscape design disciplines in that growth? Does "good design" play a role in activating a sense of "place-making" and enhancing the public perception? What schools of design and architecture in the higher education system can offer to enhance its learning environments that prepare future citizens as architects and designers?

On a larger scale, by framing the opportunity for design students to make a difference in their communities, they will learn the positive impact they can affect in the communities where they go to work. As a result, these projects will engage this community that plants the seeds for community engagement and improvement through design, in students who will work across the state, nationally and internationally. On a micro-scale, the projects will economically benefit the Brazos Valley by creating another reason to visit the downtown area to shop for fresh produce and learn more about sustainable and healthy living. It will benefit local farmers by creating a space for display, and promotion of their efforts to the public, that will serve to strengthen the value of healthy living in the community. It will benefit the local schools by providing a place where children can take field trips and be immersed in the urban agriculture through the display of wall gardens and through good design. There are doubtless many other specific beneficiaries that will be identified in the planning and programming that will occur.

Sustainable design is critically essential to sustainable living, as more people join the current movement who search for healthy and balanced living: urban agriculture can certainly benefit from good design and planning. We believe as design educators that we can make a difference in people's life through good design. Just as one might enjoy a fresh ripe and organic sweet tomato after years of tasteless produce, the effects of good design on the place-making to serve the community is priceless.

3. Rooftop agriculture

3.1. Background

Green roofs are used in urban areas to mitigate detrimental effects of urbanization. Green roofs slow down and retain precipitation, mitigate rooftop temperatures, provide habitat for regional vegetation and wildlife, and extend the life cycle of roofing materials [10–14]. Rooftops can occupy up to 35% of built land in urban environments, therefore, judicious use of flat rooftops is warranted [15]. Rooftops then are valuable space in urban areas, especially where land costs are high and structural loads allow some rooftop greening. Green roofs offer unique opportunities for entrepreneurs to produce food crops in urban areas that often lack affordable space or open ground for production [16,17]. Interest in food crop production on rooftops is growing in various climates in the United States. Some argue that production of rooftop produce is essential for food security in metro areas to meet growing global populations [1,18]. There are several outstanding examples of food production on rooftops in the United States. In New York City for example, Leslie Adatto identified 10 exemplary and popular rooftop farms in the city. At the kitchen garden scale, the Crosby Street Hotel maintains a rooftop garden that produces food for a number of items on its restaurant menu. The garden grows blueberry bushes (producing up to 15 gallons of berries in July), heirloom tomatoes, rosemary, basil, arugula and edible flowers [19]. This garden exemplifies how a rooftop garden can complement a restaurant menu. Chicago, Boston and a number of cities in the United States have restaurants with rooftop produce served on the menu.

Large-scaled rooftop farms can and do make a much more visible and widespread impact. In Brooklyn, New York, for example, the Brooklyn Grange, a one-acre rooftop, hosts one of the largest and most active open sky rooftop agriculture production sites in the United States (**Figure** 7). A project brief here outlines some of the important components and functions of rooftop agriculture. The crops on the Brooklyn Grange include vegetables, herbs and cut flowers. Crops grow in raised mounds from 20 to 25 cm high. The soil is a manufactured growing medium supplied by Rooflite®. Irrigation is provided by drip lines watered daily for 30–40 min depending upon watering needs of plants and weather conditions. There are also a few chickens raised on the roof. Environmental concerns over growing plants and food production on rooftops are valid and can only be determined through individual research. The Brooklyn Grange site has been subject to a few studies and was determined to leach slightly high levels of pH, runoff that is more turbid than a comparison sedum green roof [1]. This

Figure 7. Brooklyn Grange rooftop crops and drip tube irrigation are shown in the foreground, with the New York City skyline in the background (photo by B. Dvorak).

means that when it rains enough to generate runoff, the Brooklyn Grange roof may sometimes carry more effluent compared to the sedum roof. However, the sheer volume of stormwater retained likely outweighs the effect of occasional effluent. The roof absorbs more than a million gallons of storm water annually [19]. No urban heat island benefits or social benefits have been determined yet and would be needed to help gain a comprehensive view of environmental impacts of rooftop agriculture.

Rooftop agriculture will only maintain a presence well into the future if it is environmentally and economically sustainable. The Brooklyn Grange rooftop farm also engages the public and sustains social and cultural benefits and thus exemplifies the "Triple Bottom Line" [19,20]. The farm welcomes trainees to work on the farm and is open to visitors for self-guided tours. The roof is open to celebratory dinners for events and organizations. For example, the 2015 *Cities Alive* green roof annual conference hosted a guided tour and dinner serving a menu made from seasonal produce from the roof.

3.2. Case study of the Texas A&M University rooftop farm pilot study

Climates with long summers and mild winters have much potential to make use of rooftops to grow food; but little is known about which food crops grow on rooftops in these climates. This study investigated food crop varieties on a green roof and a living wall system for one growing season in Central Texas. The objective of this study was to evaluate the suitability of various crop species for agricultural production on extensive green roofs in Texas and to determine which species generated the greatest yield. Thirty-one crop varieties were investigated on eighty-one extensive modular green roofs. Plants were installed on the green roof and living wall during the fall of 2014 and spring of 2015 and harvested in spring and summer.

3.2.1. Green roof pilot study

The investigation site is located in College Station, Texas, on top of the Langford Building, a four-storey building at Texas A&M University (**Figure 8**). Crops tested on the green roof modules include arugula, basil, beets, broccoli, Chinese cabbage, chives, cilantro, garlic, kale, lettuce, mints, parsley, radish, shallots, spinach, strawberry, Swiss chard, thyme and turnips.

Crops were installed in eighty-one 4.5" deep modular irrigated green roof trays. The plastic trays were supplied by TectaAmerica (Corp, Skokie, IL) and each tray contained 3.5 inches of FLL compliant engineered soil used as the growing medium for extensive green roofs (Rooflite®drain, Skyland USA LLC). The trays included a non-woven geotextile used to retain moisture. Shallow retention cups used for drainage were filled with expanded shale and 12 holes located about 1 inch above the bottom of the drainage cups provide for excess water. The water holding capacity of the trays is about a 1 inch depth of rainfall. The irrigation was run about 20 min per day [20].

At the time of planting (October, 2014) the growing media was amended with 0.5 lbs of topsoil, and a half cup of Osmocote14-14-14. An additional quarter cup of Osmocote was added to each module January 30, 2015. Modules containing strawberry, lettuce, kale, spinach, broccoli, arugula and Swiss chard were fertilized using Peter's 20-20-20 fertilizer with micronutrients at a rate of one-third gallon per module and 1 teaspoon of Peter's per gallon water on March 6, 2015 [21].

Regarding the late winter early spring harvests, we found that the most productive survivors (80%+) include chives, cilantro, parsley, thyme and mint among the transplants. Among direct seeded crops, strong survivors include arugula, garlic, kale and shallots (**Table 1**).

Crop	*Variety*	*T/D*	*Percent survival*
Mint	Mojito	T	100
Mint	Peppermint	T	100
Mint	Spearmint	T	100
Thyme		T	100
Arugula		D	94
Shallots		D	93
Chives		T	92
Garlic		D	88
Kale	Beira, Red Russian, Toscano	D	86
Parsley		T	86
Beets		D	84
Strawberry	Various	T	81
Lettuce	Butterhead, Iceberg, Romaine	D	77
Broccoli		D	75
Cilantro	Chinese	T	75
Chinese cabbage		D	67
Swiss chard		D	67
Radish		D	55
Spinach	Emperor	D	53
Basil	Bell pepper	T	50
Turnips		D	9

Table 1. Percent plant survival on January 21, 2015. T = live transplant and D = direct seed.

Turnips had the least amount of production followed by spinach, radishes and basil. Many of the strawberry and mint plants died back shortly after transplanting, only to fully recover by mid-March. Strawberry plants had good survival; however, they did not produce much marketable biomass. Birds also found some of the fruits and thus the berries were not marketable. Among harvested plants, the leafy greens and parsleys were the most productive. Butterhead lettuce, romaine lettuce and Beira kale produced the greatest total amount of marketable biomass. Parsley was also a productive crop, and generated the most biomass per capita. Red Russian kale generated proportionally the most non-market biomass, as 72% of its biomass production was lost to aphids. While the lettuce varieties were nearly equal in terms of productivity, marketable biomass output was dominated by a single variety of each kale, parsley and spinach. Leafy greens, including lettuce and parsley and other herbs, were found to be viable [21].

Figure 8. The Texas A&M University green roof modules on the Langford Building with crop varieties shown here with about 3 months' growth (photo by B. Dvorak).

The crops planted late spring grew into the summer. Some of the highlights of the summer crops include several plants that had great survivability and production. Two mini-watermelons each produced one fruit of about 6 ounces each. The eight tomato plants produced about two to four tomatoes each plant, ranging from 2 to 4 ounces. Four yellow bell pepper plants produced about four small peppers each. Banana pepper plants were productive with plants producing two to five peppers each. The mint plants installed the previous fall continued growing through the summer and produced dense stands all summer. Eleven of twelve rosemary plants survived; however, they remained small.

3.2.1.1. Maintenance

Although plant survival was fair, production of fruits and harvestable material during the summer was not as productive as anticipated. Airborne weed seeds established on the green roof and competed aggressively for space. Several large garbage cans filled with weeds from the vegetable crops were hauled off the roof twice during the summer and once during the fall. Since another green roof on the same location was not regularly watered and had few to no weeds during the same time, it was presumed that the watering to maintain the crops also established and maintained weeds. We did not have weekly student activity during the summer, so maintenance was not consistent. A part-time staff working on the roof would have been very useful in keeping up with maintenance.

Daily watering of the roof was assumed to be a requirement for productivity, especially during July and August when temperatures approach 95°F and above daily, throughout the summer. The watering system was overhead spray irrigation. Since the pilot study had a large variety of plant species and varieties under the same watering treatment, some plants likely received more water than needed. We recommend future set ups to include segregating plants by water needs.

Pests affected some of the crops on the green roof. Snail shells were found on both the green roof crops and the living wall. Crop viability was influenced by position on the green roof (whether this is due to nutrient content, sunlight, irrigation, etc. should be further studied). Other crops and varieties appear indeterminate with the data at hand, and require further experimentation or harvesting to assess viability for green roof production.

3.2.2. Living walls

Seventeen crop varieties were investigated on a FloraFelt (fabric- and soil-based system) living wall (**Figure 9**). Drip irrigation was applied daily from the time of installation until early December, and from late January until the end of spring 2015, except during freezing temperatures. This system is designed to uses plants pregrown from nursery stock in containers (1 gallon size or less). Plant installation typically requires plants to be removed from containers and wrapped with a fabric blanket (FloraFelt® Root Wrap) and stuffed into a pocket of the preassembled panel. For the vegetable living wall project, some plants were purchased as small plants and then installed into a pre-existing media. Some plants were seeded. The prefabricated wall panels include a stiff backing (HDPE plastic) and a thick facade of felt. Felt is made from 100% PET recycled water bottles, is mildew and odor resistant and water absorbent. Drip irrigation tubing was supplied with each panel. Drip emitters flow a half gallon per hour and are spaced approximately three emitters per panel. Total irrigation for the wall was about 103 gallons per week during the growing season. Watering was shut off during times where overnight temperatures fell below freezing, about seven times.

Plant survival for the living wall was favorable; however, most plants were stunted and did not produce much harvestable material. Some of the successful crops include varieties of lettuce, garlic, kale, mint, peppermint, spearmint, spinach, shallots, strawberries and collards during early to middle spring; however, the irrigation system failed late spring and summer and the experiment ended. We learned that different varieties of a given crop may respond differently to the micro-climate conditions of the living wall. The water delivery system was not effective as we found that some portions of the wall remained dry and others were sufficient. Subsequently, after this study we removed the drip emitter watering system, which was out of alignment with the planting pockets, and replaced it with a soaker hose system for even distribution.

Figure 9. The living wall crops shown here after 3 months' growth. Plant survival was favorable; however, plants did not produce much fruit. Strawberry plants had high survival, but few to no berries were produced (photo by B. Dvorak).

3.3. Conclusion

This edible green roof and living wall pilot study demonstrates that some food crops may be adaptable to rooftop production in Central Texas for winter, spring and perhaps summer. Climate during the investigation was typical regarding temperature and precipitation and most plants fared well. The green roof had less competition from weeds during winter and early spring, but summer saw significant competition from weeds. We believe that such an operation demands a dedicated position to manage and maintain plants during production. Since we were not selling or distributing crops, we did not generate revenue to maintain such a position. The restaurant and large-scale case studies in New York and elsewhere had revenue to support full-time staff devoted to maintaining produce. We have plans to participate more actively with a local organic farm on campus called the Howdy Farm. We look to grow the edible rooftop projects with greater resources and participation. We believe these concepts are feasible and directly applicable to local outlets such as the Howdy Farm, farmers markets and or local production of food on campus.

References

[1] Ackerman K, et al. *Sustainable Food Systems for Future Cities: The Potential of Urban Agriculture*. 2014, 2014. 45(2, Summer): p. 18.

[2] Eigenbrod C and N Gruda, *Urban vegetable for food security in cities. A review*. Agronomy for Sustainable Development, 2015. 35(2): pp. 483–498.

[3] Texas L.o.I.V.o. *Brazos River Bottom*. 2016 [cited 2016; Available from: http://independentleaguetx.org/brazos-river-bottom/.

[4] North Texas U *The Portal to Texas History*. Brazos Co. 2016; Available from: http://texashistory.unt.edu/.

[5] Johnson R.e, RA Aussenberg, and T Cowan, *The Role of Local Food Systems in U.S. Farm Policy*, 2014, Congressional Research Service.

[6] AIA, *Farmers Market Pop-Up Project Design Competition*. 2014.

[7] Fullhart S, *Back to the Future: Farmer's Market Moving Into Downtown Bryan Area* in *KBTX*2015, KBTX: www.kbtx.com.

[8] Zawistowski K. and . M Zawistowski. *Covington Farmers Market*. 2011 [cited 2016; Available from: http://www.designbuildlab.org/gallery-2/covingtonfarmersmarket/.

[9] Creadon P, *If You Build It*, 2014: New York City. p. 1h 26m.

[10] Sutton RK, *Green Roof Ecosystems*. Vol. 223. 2015, Switzerland: Springer.

[11] Lambrinos JG, *Water Through Green Roofs*, in *Green Roof Ecosystems*. 2015, Springer. p. 81–105.

[12] Simmons MT, *Climates and Microclimates: Challenges for Extensive Green Roof Design in Hot Climates*, in *Green Roof Ecosystems*. 2015, Springer. pp. 63–80.

[13] Dvorak B, *Eco-regional Green Roof Case Studies*, in *Green Roof Ecosystems*, SuttonR.K., Editor. 2015, Springer: New York. pp. 391–421.

[14] Luckett K, *Green roof construction and maintenance*. 2009: McGraw-Hill.

[15] Peck S and J Richie *Green Roofs and the Urban Heat Island Effect*. Buildings, 2009. 4.

[16] Whittinghill LJ and DB Rowe, *The role of green roof technology in urban agriculture*. Renewable Agriculture and Food Systems. 1(1): pp. 1–9.

[17] Francis RA and J Lorimer, *Urban reconciliation ecology: The potential of living roofs and walls*. Journal of Environmental Management, 2011. 92(6): pp. 1429–1437.

[18] Peters KA, *Creating a sustainable urban agriculture revolution*. Journal of Environmental Law and Litigation, 2010. 25(1): pp. 203–247.

[19] Adatto L, *Roof Explorer's Guide 101 New York City Rooftops*. 2014, New York, New York: Leslie Adatto.

[20] Dvorak B et al., *Plant Survival for Living Walls in a Subtropical Climate*, in *Cities Alive 12th Annual Green Roof & Wall Conference* T. Cardinal Group, Editor 2014: Nashville, TN. p. 10.

[21] ZA Pace et al., *Assessing Crop Viability for Agricultural Production on Extensive Green Roofs*. in *Texas A&M University Student Research Week*. 2015. College Station, Texas: Texas A&M University.

Chapter 3

Urban Gardening: From Cost Avoidance to Profit Making — Example from Ljubljana, Slovenia

Abstract

In this study, we compare two economic models of urban gardening in Ljubljana, Slovenia. First is an avoided costs model (ACM) and the second one is a business model (BM). Comparison is made to exemplify the main economic differences between the two models. The difference is that producers under the BM sell surplus products, which is not the case under the ACM. The main aim of this study is to present an analysis of the phenomenon of urban gardening as a BM for small family home or allotment gardens. The survey was performed through Internet questionnaires and in some cases also with on-site interviews. Totally 127 urban gardeners from Ljubljana municipality participated in the research. The average ACM urban gardeners had on 1 m^2 revenue of 4.86 EUR/m^2, costs of 1.48 EUR/m^2 and gross margin (savings) of 3.38 EUR/m^2. Altogether, ACM brings savings of approximately 462.7 EUR per average size garden (136.69 m^2) or 203 EUR per median size garden (60 m^2) to the average gardener. The average BM gardener sold to the wholesale company approximately 107.0 kg of vegetables per year from 32.48 m^2 of production area for an average retail price of 1.46 EUR/kg and earning revenue of 156.44 EUR/year. Costs were approximately 21.27 EUR/year. Therefore, the gross margin or earning from surpluses sold was approximately 135.17 EUR/year for the average BM gardener, which was 4.29 EUR/m^2 or 1.26 EUR/kg of produce. The study offers evidence that the ACM can be upgraded with the BM. For example, if a family of two retired members have an average garden of 136 m^2, they can produce vegetables for four people. Consequent surpluses for two family members can be sold for extra money. The BM should be more promoted among urban gardeners as it can offer additional income and in certain cases, when a hobby becomes a profession, also a full-time job.

Keywords: urban gardening, cost avoidance, profit making, economy, vegetable production, business model

1. Introduction

Urban gardening is a food-growing concept in general covering home (backyard), allotment, community and rooftop gardens, which are taking place in the intra-urban (centre) or peri-urban (suburb) fringe of the city [1]. Urban gardening became increasingly popular in the last decade in the other parts of the world and also in Europe as a modern way of recreation and connecting with nature. Although this phenomenon is not new, it is experiencing great attention from the media as well as from policy makers and experts from various scientific disciplines.

The beginnings of urban gardens date back to Europe in the early 18th century as a response to urbanisation and industrialisation of the cities. With people immigrating at the beginning of the 19th century, this habit began to spread to other continents. At that time, the main reasons for gardens in the urban areas were mitigation of socioeconomic hardships, poverty of the working class as well as the overall weak supply of vegetables in urban areas.

The most recent "boom" in gardening is connected with solving many of the urban area's problems, which are not always related to food security but rather related to social and health (physical and mental) problems of the population, their limited access to green spaces and the economic and cultural revitalisation of degraded urban areas. However, the recent increased interest in gardening is also linked to the increasing concern of the population about food quality and costs as well as food insecurity and self-supply [2].

Urban gardening brings many different benefits, besides just growing food, which are important on personal level as well as on community level. Among them are the most important social networking and societal acceptance, recreation, health, food safety, economic cost avoidance and care for ecological issues. Urban gardening is usually considered as local food production, which enables and improves food security in the cities and contributes to its sustainable management. It also helps to diversify nutrition, development of local business, regulate poverty and social inclusion of underprivileged groups [3, 4].

In this study, we concentrate on the economic benefits of urban gardening for citizens. It is a source of self-provision found to benefit households as cost avoidance. Local residents who grow food in their backyards or in local allotment and community gardens can sell it in local markets, shops or restaurants and profit from this kind of business model (BM) [5, 6]. Certain cities support different urban gardeners groups or associations with offering them public market space as an opportunity to sell or exchange their products. Expansion of urban gardening also offers new business opportunity in producing compost from organic wastes collected at vegetable market or agro industries. Training at urban gardens is a good opportunity for young persons to evolve their environmental, agricultural and food processing skills and in the future careers [7].

On the other hand, urban gardening also contributes to the green infrastructure in the cities and is part of the provisioning ecosystem services. In geographical context, cities only produce a small share of the food they consume; however, in certain areas and time periods (political and financial crisis) this production can play an important role for food security in the city [8, 9]. Ecosystem services' benefits are often difficult to value due to multiple value parameters, which need to be taken in consideration. An interesting example for exploring benefits of integrating ecosystem services into real economies is urban gardening, a part of urban agriculture. Urban gardening is considered as a marginal economy mainly as a hobby and cost avoidance activity, although there are examples where it was turned into a profit-making activity.

Scientific literature offers us numerous examples of how urban food producers, individuals or organisations with the knowledge and desire to produce food in the city for the city market sale, succeed to overcome the obstacles to entrepreneurial initiatives [2, 4, 5, 10–12]. Obstacles can be divided into four broad categories related to site (contamination, security and vandalism, lack of long-term site tenure), government (local, federal or state government impediments), procedure (financial resources, qualified staff, time limits, small-scale projects, coordination, conflicts among partners, business planning, start-up costs) and perception of the public (food safety, low economic payback, generation gap, gardening seen as rural activity) [10]. The authors from the global North and South report on actually the same obstacles with a few decades lag time from the north to the south [2, 12–14].

Urban gardening production in the cities is in many aspects quite different from agriculture in rural areas. On one hand, numerous agricultural inputs, such as mineral fertilisers, plant protection products, irrigation water, soil medium and other, are the same. However, as urban gardening is not part of EU common agricultural policy needs on the city level special regulation and inspection. This includes specification of conditions under which this type of production is possible such as physical infrastructure, land availability and soundness with the city spatial plan, all with the purpose to expand urban gardening and avoid potential food safety risks [1, 15]. Nevertheless, cities worldwide are becoming more and more aware of the benefits of urban gardening because it can bring food security as one of the most important provisional ecosystem services. It is estimated that with potential yields of up to 50 kg/m^2 per year and more (in greenhouses), vegetable production is the most significant component of urban food production improving the food resilience of cities and contributing to global food security [15].

In this study, we compare two economic models of urban gardening in Ljubljana. First is an avoided costs model (ACM) and the second one is a BM. Comparison is made to exemplify the main economic differences between the two models. The main general difference is that producers under the BM sell surplus products, which is not the case under the ACM, where the products are consumed by producers and where surpluses are given to family members and friends for free or exchanged for other goods, rather than sold.

The main aim of this study is to present an analysis of the phenomenon of urban gardening as a BM for small family home or allotment gardens. We examine the economic benefits of urban gardening for producers beyond the food self-provision ACM and concentrate on possible entrepreneurial initiative for improving the economic status of certain social groups in the cities and for improving long-term food resilience of the cities.

2. Materials and methods

2.1. Study area

The city of Ljubljana is the capital of the Republic of Slovenia, administratively a part of the City Municipality of Ljubljana (CML) and, in broader terms, a part of the Ljubljana Urban Region (LUR) (**Figure 1**). The CML covers an area of 275 km^2 and encompasses 1.36% of Slovenian territory (20,273 km^2) and has 278,789 inhabitants, making up 13.5% of the population of Slovenia (2,062,874) [16].

The dense core of Ljubljana is integrated with other municipalities in the LUR encompassing 26 municipalities with a total of over 500,000 residents. The MOL has the highest population density 1.044 inhabitants per km^2 in Slovenia (102 per km^2), is economically one of the most developed and has the highest index of living standard among all municipalities in Slovenia. The MOL plays a key role in the entire area of the LUR and Slovenia, connecting the region and country into an integral whole with its administrative and economic power, traffic ways and daily labour migration.

In 2010, CML had 826 farms with average size of 6.9 ha. Dairy milk production is concentrated in the flatland and beef production in the hills around the city. Cereals production is constantly in decline and is subordinated to livestock and vegetable production. Fruits grown in the MOL are strawberries, blueberries and apples. Vegetable production in winter is concentrated on lamb's lettuce, rocket, lettuce and radish and in the summer months on tomatoes, potatoes, peppers, cucumbers, cabbage and lettuce. Asparagus and oil pumpkins are becoming more and more economically interesting for growing. CML systematically encourages development of integrated and organic farming. In 2009, 19 farms were included in the control of the organic farming. This type of farming is especially suitable for water protection zones in the northern part of the municipality with restriction on use of fertilisers and plant protection products.

Urban gardening is a very popular activity in the city of Ljubljana because around 10,000 inhabitants of different age groups, occupations and social status are engaged in this activity. Through history, city authorities always dedicated special areas owned by the municipality for allotments. Due to strengthening the urban gardening regulations in the past 30 years and new construction development, the number of allotment garden sites and hectares has declined from year 1984 with 200 ha, 1995 with 267 ha, 2005 with 186 ha, 2008 with 130 ha to 2014 with 158 ha. In 2010, MCL adopted a municipal spatial plan and dedicated 46 ha on 23 sites for public allotment gardens, out of which 20 ha were in production by the year 2014. Beside public areas for allotments, this is a very wide-spread private initiative in Ljubljana with home gardens and allotment gardens on hired private land. Out of 158 ha of allotment gardens, 138 ha is on private land (**Figure 1**).

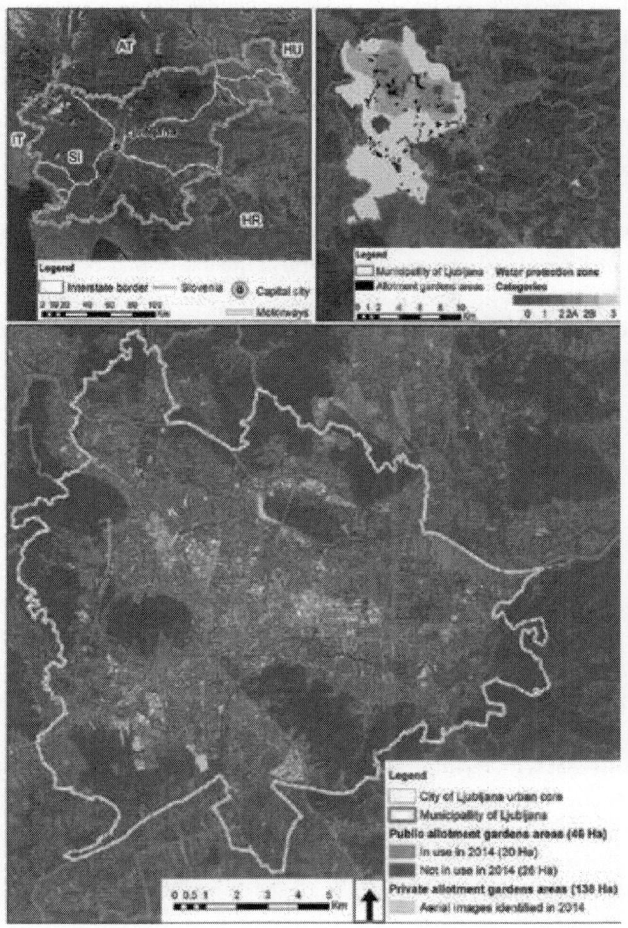

Figure 1. Study area land use by the City of Ljubljana Spatial Plan in 2014.

The city of Ljubljana started with stricter regulation in 2009 with two documents regulating allotment gardens in the ownership of the Municipality of Ljubljana. The decree on the organisation and delivery of gardens into the lease (83/2009) and Regulation for management of the allotment garden areas in the Municipality of Ljubljana (28/2009). The definition of allotment gardens and gardening in the Regulation and Decree is very strict. An allotment garden in the decree is land intended for the production of vegetables and fruits and cultivation of ornamental plants for gardeners' own purposes—self-supply. Allotment gardening is a leisure activity that involves the cultivation of vegetables and fruits and cultivation of ornamental plants with the aim of self-sufficiency and non-economical production.

Allotment gardens owned by the Municipality of Ljubljana are given in to lease to persons residing in the area of the Municipality of Ljubljana, who do not own land in the Municipality of Ljubljana, suitable for an allotment garden, and also if such land is not owned by any of the other household members. Allotment gardens can be rented for a minimum period of 1 year and a maximum of 5 years. The leasehold relationship can be extended at the request of a tenant after the expiry of the lease period for a period of 1–5 years, unless the Municipality of Ljubljana needs the land for other purposes. Only plant protection products and fertilisers allowed by regulations for organic production may be used in the garden. In gardens, which are in water protection areas, it is only permitted to grow vegetables, fruits and ornamental plants in a manner which is prescribed by the regulations in force for this area. Tenants are obliged at all times to allow sampling of soil and plants to control the use of plant protection products and fertilisers. For watering the allotment garden, rainwater collected in a uniform format storage tanks or containers is primarily used. However, all allotment sites are in general connected to the water supply system.

On the other hand, gardens on private land (home, allotment or community) which are in majority are not regulated in any way. As they represent small-scale production, they are not part of common agricultural policy subsidy payments and cross compliance to which the majority of EU environmental directives is practically attached (Nitrate and Water Directive). Small-scale urban gardening can have an impact on the quality of environment in the case of unprofessional management of production, which can be seen in pollution of soil, air, water and yield. This kind of problem could be observed when gardeners are shifting from the cost avoidance to the BM where higher yields are desired. A lack of basic agriculture knowledge and skills, which leads to unprofessional use of fertilisers and plant protection products (PPP), is especially dangerous for the environment. While legislation on the availability of PPP for non-professional use is very strict in Slovenia and they are almost unavailable for small-scale gardeners, it is different with fertilisers, which are not regulated. The results of a study on 100 allotment gardens show a high potential of possible over fertilisation of gardening plots [17]. Average soil content of phosphorus and potassium in gardening plots was 91 mg P_2O_5/100 g of soils (maximum was 366 mg) exceeding optimal values of 13–25 mg/100 g and 31 mg K_2O/100 g of soils slightly exceeding optimal values of 20–30 mg/100 g. Residues of PPP were detected only once among all samples, but concentration was below the level of detection.

2.2. Data collection

2.2.1. ACM data collection

The data in this chapter were obtained from a survey carried out in 2014 within the framework of the international project 'Food Planning and Innovation for Sustainable Metropolitan Regions' FOODMETRES (7th Framework Project, subsidised by the European Commission). A special questionnaire was developed to analyse the phenomenon of urban gardening depended on local natural resources and to examine the socio-economic benefits of urban gardening beyond the provision of food. Additionally, the aim of this questionnaire was to get an insight into specific positive and negative externalities that urban gardening is bringing to metropolitan areas. In this regard, a survey questionnaire encompasses 44 questions covering seven topics addressing different perspectives, such as (1) characteristics of grown space; (2)

growing methods applied by gardeners; (3) gardeners' skills and knowledge; (4) gardeners' motivations for gardening; (5) contribution of gardening to food supply and household budget; (6) the impacts of growing one's own food and (7) gardeners' socio-demographic characteristics. In our analysis, we took into consideration particularly the characteristic of growing space, contribution of gardening to food supply and household budget and the impact of growing one's own food.

The survey was performed through Internet questionnaires and in some cases also with on-site interviews. Participants were informed about the questionnaire by e-mails with the help of allotment gardeners' associations. Overall, 127 urban gardeners from Ljubljana municipality participated in the research. The reach window lasted from 6 January 2014 to 31 January 2015. Gardeners were categorised by four garden types. Gardeners with home gardens in their backyards (55) with 43.3% represent the majority of sample. Allotment gardeners with gardens on public land and with legal contracts (34) represent 26.8% of the sample. Allotment gardeners with gardens on private land with a contract (29) represent 22.8% of the sample. Allotment gardeners on private or public land without any contract (3) represent 2.4% of the sample population. The last group with 4.7% represent gardeners with gardens as part of a community garden, with mixed ownership or balcony gardening (6).

For the economic evaluation of the ACM model, we used data on growing space area (m^2), crop varieties and harvested yield (kg), total costs for production (euro) and surplus vegetable management. Additionally, we also used labour invested in management of gardens (hours). All data were self-reported by participants. The data on produce harvested were the most time demanding and challenging for the gardeners to report because they usually do not measure yields or areas of growing spaces. For this matter, we developed a questionnaire where they were able to report about their vegetable production in three different ways: as harvested yield in kilograms (e.g. 1 kg of onions) or as surface area of production in square metres (e.g. 1 m^2 of onions) or as number of individual plants of produced vegetable (e.g. five plants of onions). If any of the values was missing, it was later converted with transfer tables developed specially for this research (**Table 1**). For example, 1 kg of onions is equivalent to 0.2 m^2 of land or 10 individual plants of onions.

This enabled us to calculate average harvest yields for each of the different types of vegetables grown and also a single harvest value in kilograms per square metre made up of a weighted average of vegetables grown in the average size garden.

Vegetable type	Harvested yield (kg)	Area (m^2)	Number of plants or amount of seed for 1 kg of yield
Beans green	1	0.6	25
Broccoli	1	0.5	2
Cabbage	1	0.4	1
Carrot	1	0.15	15

Vegetable type	Harvested yield (kg)	Area (m^2)	Number of plants or amount of seed for 1 kg of yield
Cauliflower	1	0.33	1.33
Celery	1	0.2	2
Cherry tomato	1	0.15	0.4
Cucumber (for pickling)	1	0.25	1
Cucumber salad	1	0.125	0.5
Egg plant	1	0.25	1
Garlic	1	1.4	28
Green cabbage	1	0.25	1
Kohlrabi	1	0.25	5
Leek	1	0.2	6
Kale	1	0.25	4
Corn salad	1	1	80
Onions	1	0.2	10
Paprika	1	0.25	1
Parsley	1	0.33	33
Peas	1	1.33	130
Pepperoni	1	0.25	1
Potatoes	1	0.33	2
Pumpkin	1	0.5	0.5
Radicchio	1	0.5	4
Red Beet	1	0.25	12
Salad	1	0.25	3
Tomato	1	0.20	0.7
Turnip	1	0.25	10
Zucchini	1	0.25	0.25

Table 1. Vegetable conversion table between yield (kg), area of production (m^2) and number of plants or seeds for 1 kg of yield—for most common vegetables grown in Ljubljana.

2.2.2. BM data collection

The data for the BM model were obtained from FOODMETRES SME project partner. The partner wishes to stay anonymous, therefore, we refer to it as the Company. The Company is a medium-size private company with a 12% share in distribution of vegetables and fruits in Slovenia and with an annual turnover of 18 million EUR. It deals with the import and export of vegetables and fruits. In an average year, the Company purchase food products from 160 individual producers (professional farmers) and from 5 cooperatives from Slovenia and from 30 urban gardeners from Ljubljana. The main reason for buying vegetables from small urban gardeners in Ljubljana is customer demand for locally produced food and local varieties of vegetables (e.g. Ljubljanska ledenka/Ljubljana iceberg lettuce). This way, the Company has access to products that are not produced in mainstream agricultural production, but are highly valued on the market.

The data were obtained from an interview with a representative of the company that buys and sells produce from urban gardeners in Ljubljana and their database. Data for the years 2014 and 2015 were used to ascertain how many urban gardeners are selling their products to the Company, what the total amount of vegetables sold by urban gardeners was and how much an average gardener earned per year.

2.2.3. Calculation of savings and earnings

In the first step of calculating savings of ACM, we calculated the area of an average garden from the data gathered with questionnaires. In the second step, we used the supplied data from the questionnaires on vegetable varieties grown in an average garden (kg, m^2). We extracted the 15 most grown vegetables by an area in an average garden and calculated the share of each. Then, we calculated the average annual harvested yield (kg/m^2) for each of the most common vegetables. This enabled us to multiply the share and average yield of the most common vegetables and obtain harvested yield (kg/m^2) for an average garden. In the third step, we obtained average retail prices (EUR/kg) for the most common vegetables from the Statistical Office of the Republic of Slovenia for the year 2014, as correspondents reported production for that year. In the fourth step, we calculated revenue, cost and gross margin for average garden size. Revenue for an average garden (EUR/m^2) was calculated by multiplying yield (kg/m^2 average garden) and retail price (kg). Costs of gardening production were gathered from questionnaires as total cost for garden management per year, including seeds, seedlings, fertilisers and plant protection products. Costs per kilogram of produced vegetable (EUR/kg) were also calculated. The gross margin in Euros per average garden was calculated as total revenue (EUR) multiplied by total cost (EUR). The fifth step was recalculation of revenue, cost and gross margin numbers to arbitrary areas of 1 m^2, 100 m^2 and 1 ha, municipality designated area for gardening (46 ha) or total gardening area observed from aerial images (158 ha).

For calculating earnings of BM, we took the total bought vegetables (kg, EUR) from the urban gardeners and divided it by the total number of urban gardener suppliers to the Company. The result was average yield sold and average earning per individual gardener.

3. Results and discussion

3.1. Average garden size in Ljubljana

Based on the survey results, it was determined that the average urban garden size in the Ljubljana is 136.69 m^2 (**Table 2**). However, the amount of land used for vegetable production varied considerably. The data collected by questionnaires show that garden size ranged from approximately 1 m^2 to 5000 m^2. An average is not always useful for representation of a central tendency so we therefore calculate a median of 60 m^2 which is a better estimation of the usual garden size. The city of Ljubljana offers its citizens a uniform public urban garden of exactly this size.

Using ArcGIS, we determined from the municipality spatial plan that the city of Ljubljana has designated 46 ha of urban gardening areas for allotments. From aerial orthophoto images, we determined that the city currently has 158 ha of allotment garden areas of which 20 ha is publicly owned and 138 ha privately owned. Based on these numbers, lowered for 10% of functional land (paths and garden sheds), we can estimate that there are approximately between 10,000 (average size) and 23,000 (median size) allotment gardens within the city borders.

Type of gardeners	Gardens		Area (m^2)	
	Number	Percent (%)	Sum	Average
Gardeners with home gardens at their backyards	55	43.31	10,753	195.51
Allotment gardeners with garden on public land and with legal contract	34	26.77	1797	52.85
Allotment gardeners with garden on private land with contract	29	22.83	4080	140.69
Allotment gardeners on private or public land without any contract	3	2.36	90	30.00
Gardeners with gardens as part of community garden, with mixed ownership or balcony gardening	6	4.72	640	106.67
Total	**127**	**100.00**	**17,360**	**136.69**

Table 2. Number and area (m^2) of gardens by gardener's type included in the study for Ljubljana in 2014.

We estimated that the use of average garden size (136.69 m^2) for economic calculation is the best value because in further calculations we also used averages for production yield, costs, saving and earning.

3.2. Crop types in production

All urban gardeners in the survey report on growing 43 different species, sorts or varieties of vegetables. The average gardener grows 15 different crops (**Figure 2**). More than 15 types were grown by 44.5% or 39 gardeners. The maximum number of vegetable types was 32, grown by three gardeners. All surveyed gardens together harvested 15,711 kg of vegetable in 2014. Out of which the 15 most common crops represent 78.6% or 12,348 kg of harvested vegetables.

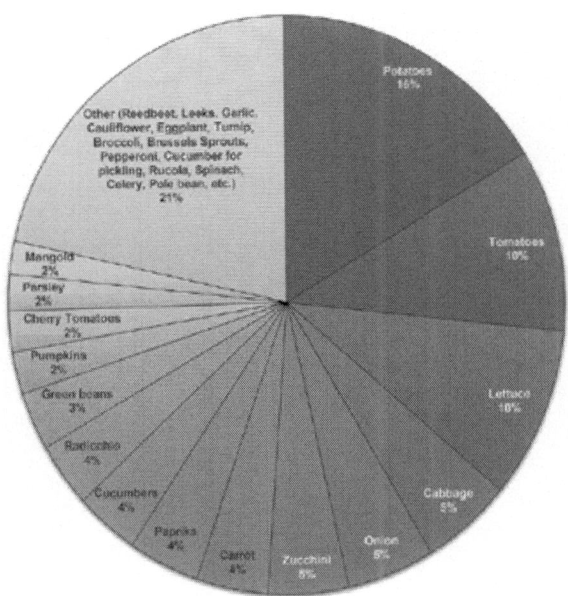

Figure 2. Shares (%) of the crops (kilograms harvested) across surveyed gardens.

The most grown crop in surveyed urban gardens is potatoes with 45.2 kg grown on 15.9 m^2 of the average garden area, representing 16.4% (2575 kg) of the total harvest (15,711 kg). Tomatoes were the next most common crop with 19.8 kg on 5.9 m^2 of area representing 10.3% of the total harvest. Third, lettuces (leafy greens) were with 16.7 kg on 6.3 m^2 of area representing 9.5% of the total harvest in the reported year. Shares of other grown crops based on harvested yields in the surveyed gardens are shown in **Figure 2**. In further economic calculations, we use only the data for the 15 most represented crops as a proportion (%) of weighted harvested yields (potatoes, tomatoes, lettuce, cabbage, onion, zucchini, carrot, paprika, cucumber and radicchio). This enables us to exclude some marginal crops and uncertainties in correspondents' reporting.

3.3. Production yield

The results of the survey showed that 127 gardeners produced an average of 2.16 kg of yield per m^2 with a standard deviation of 1.51 kg/m^2 and the range of 7.48 kg/m^2. Statistics shows considerable variation among gardening respondents. The top 10%, top 20% and top 30% of the respondents harvested on average 5.58 kg/m^2, 4.57 kg/m^2 and 4.09 kg/m^2, respectively. Almost 35% or 44 gardeners produce more than the average. This group of gardeners is very productive and shows that high yields are possible also in small urban backyard or allotment gardens. This could be connected to the fact that the average gardener has 17 years of gardening experience (minimum 1 and maximum 60 years).

The most grown 15 crop varieties in average garden	Area in average garden		Yield in average garden	
	(%)	(m^2)	(kg/m^2)	(kg)
Potatoes	20.86	28.51	3.52	100.35
Tomatoes	13.16	17.99	3.87	69.61
Lettuce (leafy greens)	12.14	16.59	2.98	49.45
Cabbage	6.65	9.09	3.50	31.83
Onion	6.48	8.86	2.96	26.21
Zucchini	5.93	8.10	3.93	31.85
Carrot	5.30	7.24	3.85	27.87
Paprika	5.26	7.20	2.90	20.87
Cucumbers	4.71	6.43	5.70	36.66
Radicchio	4.55	6.22	2.09	12.99
Green beans	4.10	5.60	1.94	10.86
Pumpkin	2.96	4.04	3.28	13.25
Cherry tomatoes	2.90	3.97	4.05	16.08
Parsley	2.63	3.60	2.00	7.19
Mangold (Chard)	2.38	3.25	2.92	9.50
Sum	**100**	-	**3.40**	-
Average size garden	-	**136.69**	-	**464.59**
Median size garden	-	**60.00**	-	**203.93**

Table 3. Areas of production (%, m^2) and yields (kg, kg/m^2) for the most common 15 crop varieties grown in average garden.

In a further calculation, we considered only the most common top 15 crops with more than 2% of share of the harvested yield. The average production among sampled respondents was 3.40 kg/m^2 (**Table 3**). This means that a gardener with an average-sized garden produced 465 kg of vegetables in the year 2014. A gardener with a medium-sized garden (60 m^2) produced 204 kg of vegetables.

3.4. Cost of production

Gardeners participating in this study spend on average 75.13 EUR per garden and 1.48 EUR/m^2 of garden area. If we multiply the last value by the average garden size of 136.69 m^2, we get a cost of 202.30 EUR for the average garden size in the study. Costs per garden varied widely from a minimum of 5 EUR/m^2 to a maximum of 360 EUR/m^2. These values represent only variable costs, such as buying fertilisers, seeds, seedling plants, plant protection products and small equipment. Rent for the garden plot, water and labour is not included in this value. Respondents were also asked to estimate the active time spent in the garden. Labour is estimated below. If we extrapolate the average cost per m^2 to total allotment gardening area (158 ha) in Ljubljana, not including home backyard gardens, results show that gardeners in the city spent approximately 2.3 million EUR on gardening material during the growing season of 2014.

The labour involved in maintaining the garden plots varied widely. On average, they spend 0.134 hours/m^2 per week actively working and keeping crops, which is 7.8 minutes/m^2 per week with standard deviation of 0.130 hour/m^2 per week, respectively. As the vegetation period in Ljubljana lasts from April to October (7 months or 28 weeks), we can estimate for the growing period of 2014 that the average gardener invested 3.752 hour/m^2 or 512.8 hours per average garden of 136.69 m^2. Using the areas identified as allotment gardens (158 ha) in the city of Ljubljana, these results show that gardeners invested approximately a total of 5.9 million of labour hours in the growing period of 2014. As agricultural calculation models as well as gardeners do not account labour hours as costs, they were not included in the final calculation of savings as an actual cost but rather as an informative result for calculating labour efficiency.

3.5. Savings with ACM

When we express cost in EUR/kg of vegetable produced by surveyed gardeners, our results show that each kilogram costs 0.44 EUR. This is one third of the price than what the same vegetable would have cost in local grocery stores in 2014. Retail prices of the 15 most common vegetables (from potatoes to mangold) were obtained from Statistical Office of the Republic of Slovenia (2015) and calculated as a price for 1 kilogram of mixed vegetables with the same shares as presented in **Table 3**. The price of a kilogram of vegetables ranged from 0.67 EUR/kg for potatoes to 2.22 EUR/kg for paprika with average of 1.49 EUR/kg (**Table 4**). This shows that average urban gardener in Ljubljana avoids costs and saves 70% of what he would pay in stores by producing his own vegetables.

The ACM urban gardeners produced 465 kg of vegetables in an average-sized garden in 2014 with an average retail price of 1.49 EUR/kg, which resulted in average revenue of 664.7 EUR (**Table 4**). The costs for production in an average garden were 202.3 EUR. This finally resulted in the total average annual gross margin of 462.7 EUR (**Table 5**). If we recalculate this to 1 m^2, average revenue was 4.86 EUR/m^2, costs were 1.48 EUR/m^2 and gross margin was 3.38

EUR/m². Gross margin can be understood as avoided costs or savings for the gardener. Using these values with the city designated area for allotment gardens (45 ha) and total allotment garden areas identified from above (158 ha), results suggest that the allotment gardeners of Ljubljana avoided approximately 1.5 million and 5.3 million EUR costs annually, respectively. These values could be lowered by approximately 10%, as part of these allotment areas is also used as functional lands for walking paths and garden sheds. Taking this into account in the final calculation, the avoided cost would amount to approximately 1.4 million EUR for the total public allotment area and 4.8 million EUR for the total allotment area in the city of Ljubljana for the year 2014.

Average 15 crop varieties grown in average garden	Yield (kg)	Retail price (€/kg)	Revenue (€/year)
Potato	100.35	0.67	67.23
Tomato	69.61	1.97	137.14
Salad	49.45	1.82	90.00
Cabbage	31.83	0.58	18.46
Onion	26.21	0.91	23.86
Zucchini	31.85	1.25	39.81
Carrot	27.87	1.08	30.10
Paprika	20.87	2.22	46.33
Cucumber	36.66	1.72	63.06
Radicchio	12.99	2.16	28.06
Green beans	10.86	3.95	42.92
Pumpkin	13.25	1	13.25
Cherry tomato	16.08	1.97	31.67
Parsley	7.19	2	14.39
Mangold (Chard)	9.50	1.94	18.44
Sum	**464.59**	-	**664.71**
Average	-	**1.49**	-

Table 4. Calculation of retail price and revenues for average garden.

When taken all together, we can state that ACM brings to the average gardener savings of approximately 3.38 EUR/m² or 462.7 EUR per average size garden (136.69 m²) or 203 EUR per median size garden (60 m²).

The results of the survey also show that gardeners involved a lot of hours of their own work in managing gardens and crops (**Table 5**). Based on the calculation of production and labour hours, we were able to define the labour efficiency or productivity of an average gardener. Average production and labour invested in urban gardens reported by survey respondents were 3.39 and 3.75 hours/m^2, respectively. One labour hour for agricultural seasonal work in Slovenia was valued to range from 2.2 to 5 EUR/hour approximately in 2014, with an average of 3.5 EUR/hour. An average labour efficiency was calculated at 0.89 kg/hour per m^2 of garden. Further calculations suggest that work on 1 m^2 during 28 weeks (April-October) of growing period was worth 13.13 EUR. If we would also factor in the value of labour to the calculation of gross margin, the cost of self-provisioning for urban gardeners would rise remarkably to the level when it would no longer be sustainable.

Calculation	Urban gardening area					
	1 m^2	100 m^2	136.69 m^2	1 ha	45.89 ha	158.06 ha
Production (kg)	3.39	339.9	464.6	33,989	1,559,755	5,372,165
Revenue (EUR/year)	4.86	486.3	664.7	48,629	2,231,591	7,686,128
Costs (EUR/year)	1.48	148.0	202.3	14,800	679,172	2,339,229
Gross margin (EUR/year)	**3.38**	**338.3**	**462.4**	**33,829**	**1,552,419**	**5,346,899**
*Gross margin reduced for 10% of area (EUR/year)	-	-	-	-	1,397,177	4,812,209
Labour (hours/year)	3.75	375.2	512.8	37,520	1,721,793	5.930.261
†Labour (EUR/year)	**13.13**	**1313.2**	**1794.8**	**131,250**	**6,023,063**	**20,744,850**
*Labour (EUR/year)	-	-	-	-	5,420,756	18,670,365
Labour efficiency (kg/hour)	0.89	-	-	-	-	-

*Allotment garden area used as functional lands for walking paths and garden sheds where actual production of vegetable is not possible.
†Labour hour for seasonal works in agriculture was valued with an average of 3.5 EUR/hour in 2014.

Table 5. Calculation of avoided costs model (ACM) gross margin for different areas of vegetable production in urban gardens in the city of Ljubljana for the year 2014.

3.6. Earnings with BM

BM urban gardeners sell surplus produce to obtain an additional source of income, which is not the case of ACM urban gardening for self-supply where the produce is consumed and where surplus produce is given to members of family and friends or exchanged for other goods rather than sold. The Company representative reported that BM urban gardeners are mostly highly educated people of which half are retired and half are employed. It often happens that retired urban gardeners transfer their cooperation with the Company to the younger generation (their children). The younger they are, the more willing they are to use the opportunity, although it is typical for them to have less experience, less skills and less knowledge about gardening. The cooperation of the Company with the urban gardeners sometimes results in urban farmers becoming real farmers. A BM urban gardener rents or buys agricultural land and starts with the registered production of fruit and vegetables as agricultural holdings, from which the Company buys produce. Urban gardeners sell produce produced within 1–3 km radius from the Company. They have no contract agreement because gardeners bring produce whenever they have surplus. Price is negotiated based on the market supply and demand with a Company representative. Sometimes, the Company calls reliable regular urban gardener suppliers in advance to provide them with specific local varieties of vegetables like Ljubljana iceberg or Ljubljana bog green beans.

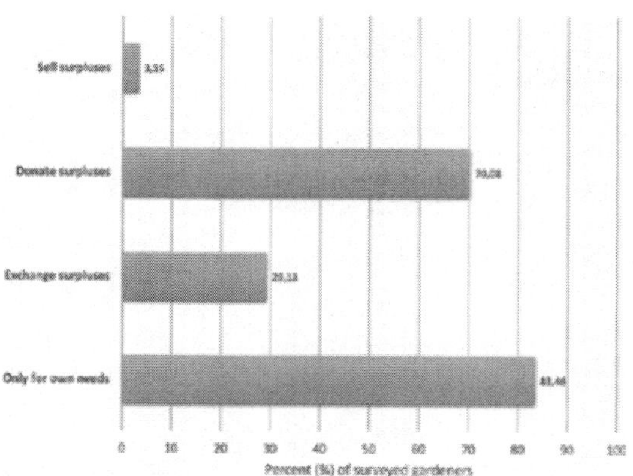

Figure 3. Paths of vegetable surplus production of surveyed urban gardeners in Ljubljana in the year 2014.

In the period 2014–2015, the Company purchased on average approximately 3209 kg of vegetables per year from 30 gardeners in the amount of 4301 EUR for an average retail price of 1.46 EUR/kg. The BM gardeners (on average 30 per year) have sold on average 107.0 kg of produced vegetable surpluses, and thus earned (revenue from surplus) 156.44 EUR/year. Results of the urban gardener's survey showed that only four gardeners (3.15%) are selling their produce (**Figure 3**). In further calculations, we made two assumptions, production and cost per square area of urban garden are the same as for ACM. Dividing the average amount of sold vegetable (107.0 kg) by average production (3.39 kg/m^2), we got 31.48 m^2 of garden area needed for BM production. Multiplying this value with average costs of production (1.48 kg/m^2) in an ACM urban garden, we got the total cost. The cost for 107.0 kg of produce amounted to approximately 21.27 EUR/year. Therefore, the gross margin or earning from surpluses sold (revenue minus cost) was approximately 135.17 EUR/year for an average BM gardener, which was 4.29 EUR/m^2 or 1.26 EUR/kg of produce. Higher earnings of BM in comparison with ACM are expected because they sell special local varieties with higher added value.

Urban gardening is not subject to the traceability of goods and quality control (certification rules) in comparison with classical agriculture. As soon as an agriculture holder enters with their products into the real economy, he is obliged to follow the certification and quality control rules. All produce supplied to the Company is examined by their food technologist, and then the quality analysis of the products follows before they enter the official food chain and market. The financial cost of one quality analysis is approximately 100–500 EUR, which normally exceeds what urban growers are willing or can afford to spend. For this reason, the Company covers all the cost of food quality control.

4. Conclusions

4.1. Strengths

This is one of the rare agro-economic evaluations of the two urban gardener economic models. This study illustrates how much ACM gardeners save with self-provision gardening and how much BM gardeners earn with selling surpluses of their production. The sample of surveyed respondents filling out the on-line questionnaire was relatively wide in number of correspondents allowing us to make a good comparison between urban gardeners and their behaviour. One of the important strengths of this research was that the growing season covered in the study passed without any serious damaging weather events (drought, hail), which could cause a loss in productivity.

4.2. Limitations

This study has some limitations impacting on the reliability of the results. The first is that only one city was studied so it would be difficult to extrapolate results beyond the city of Ljubljana. However, we could say that results are transferable to other cities in Slovenia with continental climate as weather conditions, and prices of vegetable and cost are even all around the country with a population of 2 million. The part of the country with a Mediterranean climate is an exception because the growing period for certain crops (lettuce, leeks, radicchio, etc.) lasts all year round.

Although the sample of participating gardeners in the research was relatively wide, the selection of participants was random with no fixed pattern of selection. The total number of gardeners starting filling out the online survey was greater, however due to the length and data requirements, many of them quit before finishing and were finally deleted from the database. Further, all data in the study were self-reported by gardeners based on their memories for the last growing season. For that reason, it could have happened that results would not be representative for gardeners in Ljubljana. However, as we were aware of the limitations of the on-line questionnaires, we also performed random field surveys at urban garden allotment areas with the same form of questionnaire as for on-line participants. The next limitation was connected to the BM data. According to our knowledge, the Company involved in the research is the only one among the wholesale companies in the region with a BM involving buying vegetable and fruit production from small home or allotment urban gardeners. This type of model requires the company to invest more time, financial resources and labour and is for that reason unattractive to businesses because they do not profit enough from this relationship.

Another limitation was that during the course of the research we were not able to make any contact with small urban gardeners producers that sell to the Company. The managing director of the Company helped us on that with personal involvement, however with no success. BM urban gardeners main reasons to decline an interview were concern and fear of possible subsequent visitation by the agricultural, food or even tax inspectors. They believe that their way of production is prohibited and count themselves in the grey zone of the economy.

4.3. Knowledge acquired

The study showed that gardeners were growing a wide variety of crops, on average 15 different types, and this required a lot of different knowledge and production skills. It also means that the diet of an average gardener was quite diverse. On a yearly basis, the average gardener (136.69 m^2) produced 465 kg of total vegetables or 364 kg without potatoes.

The Slovenian Ministry for agriculture, forestry and food, following World Health Organisation (WHO) recommendations, advises for a grown person to eat a minimum of 250 g of vegetable per day without potatoes. This means that the vegetable quantity produced by an average urban gardener is sufficient for 1457 daily servings, which was enough for a family of four members to be self-sufficient in vegetables for a whole year. If we compare this to a median garden (60 m^2) in the study, they produce 160 kg of vegetable without potatoes, which account for 640 daily servings of vegetable or 1.8 family members per year. The results offer us an evidence of high potential for families to be self-sufficient. There is also plenty of room and opportunity to expand the city of Ljubljana's self-provision of vegetables.

The study offers evidence that the ACM can be upgraded with the BM. For example, if a family of two retired members has an average garden of 136 m^2, they can produce vegetables for four people. Consequently, the surpluses for two extra family members can be sold for extra money.

In present times, the BM is quite rare in Ljubljana, however, in the past centuries before population growth after the World War II, this type of vegetable production was very common in Ljubljana. Housewives, living in small houses on the edge of the city wall, sold their vegetable produce from home gardens to the citizens on the main market every morning. This was additional money for their household budget, and it also meant financial independence for women. The earnings are not great but it seems that people are doing it for a two combined motives as additional income whereas food is not wasted.

4.4. Future of urban gardening

The long history of urban gardening in Ljubljana is the best guarantee for a bright future. Urban gardening is currently a subject of many social and aesthetic studies dealing with the involvement of different social groups in co-shaping the city's green areas. The municipality is spatially arranging urban gardening allotment areas with an accelerating speed putting in to use few hundred new garden plots each year. While cooperation with ACM urban gardeners was open and correct and they were always available to help us, this was not the case with BM gardeners. In future studies, a great deal of work will have to be invested in strengthening the trust between BM producers and researchers to convince them to cooperate and provide them necessary anonymity.

The BM should be promoted more because it can offer additional income and in certain cases when a hobby becomes a profession, also a full-time job. However, currently gardeners who hire gardens from the municipality are in a worse position, as municipality regulations for urban gardens on public places require only non-profit gardening for self-supply of the family. The municipality will have to reconsider its strict regulations in the future. However, urban gardeners are always one step ahead of the authorities. In the light of the lack of public urban garden plots, citizens hire privately owned fields from farmers and divide them into smaller plots. Currently, areas of this type represent 87% (138 ha) of all urban allotment garden areas (158 ha), showing that the municipality will never be able to fulfil the demand for garden plots by citizens.

This shows that the municipality of Ljubljana will have to concentrate also on possible entrepreneurial initiatives originating from the urban food chain (production and sale), which could improve self-provision and the long-term food resilience of the city.

References

[1] Mougeot LJ. Urban agriculture: definition, presence, potentials and risks. Growing cities, growing food: Urban agriculture on the policy agenda. 2000:1–42.

[2] Orsini F, Kahane R, Nono-Womdim R, Gianquinto G. Urban agriculture in the developing world: a review. Agron Sustain Dev [Review]. 2013;33(4):695–720.

[3] Jehlicka P, Kostelecky T, Smith J. Food self-provisioning in Czechia: beyond coping strategy of the poor: a response to Alber and Kohler's 'Informal Food Production in the Enlarged European Union' (2008). Soc Indic Res. 2013;111(1):219–234.

[4] Smith J, Jehlicka P. Quiet sustainability: fertile lessons from Europe's productive gardeners. J Rural Stud. 2013;32:148–57.

[5] Ranasinghe TT. A novel living agricultural concept in urban communities: Family Business Garden. Int J Sustain Dev World Ecol [Article]. 2003;10(3):239–245.

[6] Anastasiou A, de Valenca A, Amare E, Montes de Oca G, Widyaningrum I, Bokhorst K, et al. The Role of Urban Agriculture in Urban Organic Waste Management in The Hague - Academic Consultancy Training. The Netherlands, Hague: Wageningen University; 2014; 70. Available from: http://www.foodmetres.eu/foodmetres-urban-organic-waste-management-in-the-hague-report/

[7] Corrigan MP. Growing what you eat: developing community gardens in Baltimore, Maryland. Appl Geogr. 2011;31(4):1232–1241.

[8] Gómez-Baggethun E, Barton DN. Classifying and valuing ecosystem services for urban planning. Ecol Econ. 2013;86:235–245.

[9] Gómez-Baggethun E, Gren Å, Barton D, Langemeyer J, McPhearson T, O'Farrell P, et al. Urban Ecosystem Services. In: Elmqvist T, Fragkias M, Goodness J, Güneralp B, Marcotullio PJ, McDonald RI, et al., editors. Urbanization, Biodiversity and Ecosystem Services: Challenges and Opportunities. Netherlands: Springer; 2013; 175–251.

[10] Kaufman J, Bailkey M. Farming Inside Cities: Entrepreneurial Urban Agriculture in the United States. Massachusetts, Cambridge: Lincoln Institute of Land Policy200. Report No.: WP00JK1.

[11] Pfeiffer A, Silva E, Colquhoun J. Innovation in urban agricultural practices: Responding to diverse production environments. Renew Agr Food Syst [Article]. 2015;30(1):79–91.

[12] Mok HF, Williamson VG, Grove JR, Burry K, Barker SF, Hamilton AJ. Strawberry fields forever? Urban agriculture in developed countries: a review. Agron Sustain Dev [Review]. 2014;34(1):21–43.

[13] Hamilton AJ, Burry K, Mok HF, Barker SF, Grove JR, Williamson VG. Give peas a chance? Urban agriculture in developing countries. A review. Agron Sustain Dev [Review]. 2014;34(1):45–73.

[14] Poulsen MN, McNab PR, Clayton ML, Neff RA. A systematic review of urban agriculture and food security impacts in low-income countries. Food Policy [Review]. 2015;55:131–46.

[15] Eigenbrod C, Gruda N. Urban vegetable for food security in cities. A review. Agron Sustain Dev [Review]. 2015;35(2):483–98.

[16] SURS. SI-Stat Data Portal. Slovenia, Ljubljana: Statistical Office of the Republic of Slovenia; 2015. Available from: http://pxweb.stat.si/pxweb/dialog/statfile1.asp.

[17] Jamnik B, Smrekar A, Vrščaj B, Kladnik D, Erhartič B, Komac B, et al. Vrtičkarstvo v Ljubljani/Allotment gardening in Ljubljana. Ljubljana: Založba ZRC; 2009.

Chapter 4

Comparison of the Land Uses and Sustainable Development in Small Islands: The Case of Skiathos Island, Greece

Abstract

The Island of Skiathos occupies a total area of 50 km^2, accounting for 1.6% of the area of the prefecture of Magnesia and 0.28% of the Region of Thessaly, Greece. The land is hilly and can be divided into farmland, meadows, woodlands, land covered by water and land occupied by settlements and roads. Also, a large part is occupied by burnt areas that resulted from the fire of 2007. The aim of this chapter is to present the evolution of existing land uses at the Island of Skiathos during the past decades. With the contribution of Geographic Information Systems (GIS) and the orthophotomaps, the spatial planning of the land uses can be evaluated for all these years and the total area can also be calculated. Our results are important for understanding the impacts of land uses on ecosystems in the frame of sustainable development. There has been no other research regarding land uses in Skiathos Island in the past, and, also, this is the first digitization of the area. Finally, two sustainable spatial development scenarios for the Island of Skiathos are proposed. The first scenario relates to the results obtained from a prediction (application of the model of automatic cellular) while the second scenario refers to a more realistic model of development with focus on environmental protection and sustainable development.

Keywords: GIS, land uses, area, island, thematic maps, scenario

1. Introduction

Over the past few decades, European ecosystems have changed substantially as a result of socio-economic and political changes, while future transformations are also expected to occur [9,13].

The European Mediterranean landscapes have undergone many changes over the years due to the relocation of people in coastal areas, forest fires, desertification, logging, the rapid expansion of activities related to tourism and the intensification of agriculture [6,14,20].

Environmental and ecological consequences of landscape transformation are more evident in natural ecosystems where their sustainability, multifunctional role and values are threatened [10,11,15,19].

Sustainable development refers to economic development that is planned and implemented, taking into account environmental protection and sustainability.

The rule of sustainability is the maximum gain of goods from the environment, without interrupting the natural production of these products in sufficient quantities in the future. Sustainable development implies development of the productive structure of the economy while creating infrastructure for a sensitive attitude toward the natural environment and ecological problems (such as defining traditional sciences like geography).

Sustainability implies that natural resources are exploited at a rate lower than that to which renewed, otherwise occurring environmental degradation. In theory, the long-term effects of environmental degradation are the inability of the earth's ecosystem to support human life (ecological crisis).

- The coastal and insular nature of the geography of Greece has a significant impact on local development; it creates geographical areas heavily dependent on the mainland. Although the geographical distances separating the islands from mainland Greece are not large (following the natural-geographic mosaic recommended the Greek area), developmental and political choices are affecting their real integration with criteria of cost, time, frequency of service, availability of interconnection etc., creating, as a consequence, significant dysfunctions in their development completion.

Exploring the development process in the islands and a strategic integrated development model for small islands is on systematic research field and policy in world affairs for decades. The structural problems flowing from the specific natural and socio-economic characteristics distinguish the islands from the mainland and intensified in inverse proportion to the size of the islands, slowing the growth process. Although the need for a differentiated approach to the island territory and particularly small islands has been recognized by the international community in Greece, a few steps in that direction have been made. Given the insular nature of the country, the small size of the majority of the islands and the strong development catching up, exploring the roadblocks and the formation of a strategic model of integrated development of small islands space become necessary.

The purpose of this article is to apply the changes of land uses for the past few decades for the Island of Skiathos and also make a prediction for the land uses change for the year 2020. The comparison of the land uses has been made with the tool of GIS and orthophotomaps. Moreover, to make the prediction, we used the model of cellular automata and two indicators.

Our results are important for understanding the impacts of land uses on ecosystems in the frame of sustainable development. There has been no other research regarding land uses in Skiathos Island in the past, and, also, this is the first digitization of the area.

Finally, in the future, these results will be used for making the spatial plan for the Island of Skiathos.

2. Methodology

2.1. Study area

The Island of Skiathos resides in the territorial unity of the northern Sporades and the capital is the city of Skiathos. It is 2.4 miles from the coast of southern Pelion and 4 miles from Skopelos. The surface area is about 49,89 km^2 while the length is 12 km and the width is 9 km. Within the administrative boundaries of Skiathos island lie the smaller islands of Tsougrias, Small Tsougrias, Repio, Aspronisi, Maragos and Arcos. The town of Skiathos is connected with the Island of Skopelos, Alonissos, the town of Volos, Ag. Konstantinos and Euboea, via a (trade—passenger) port. There is also a National Airport located in the northeastern part of the island. Administratively, the Municipality of Skiathos is part of the Magnesia Municipality in Thessaly.

A large part of the island is covered by woodland and the rest of the island is dominated by olive trees. In Skiathos there is one area identified as the Area of Community Interest (Sites of Community Interest) and was included in the European ecological network Natura 2000 (Directive 92/43/EEC). That area is the island of Skiathos (code: GR1430003, 32 ha). Around the island there are about 70 beaches.

The population of the island, in 2011, was 6088 and the density was 122 people/km^2. In Table **1**, the population and the population density, during the past few decades, are presented.

	Population	Population Density (people/km^2)
1991	5096	102.14
2001	6160	123.47
2011	6110	122.45

Table 1. Population.

It can be observed that the difference between the last two decades is very small. The most significant increase in the population took place during the period 1991–2001.

Mediterranean climate prevails in Skiathos with cold winters and pleasant summers. There is almost complete absence of rainfall during the summer months. The temperature during July and August often exceeds 30 degrees Celsius, while in June the weather is somewhat cooler.

As in most islands, from July to mid-August we encounter the phenomenon of Meltemi, quite throwing the night temperature.

In the Island of Skiathos, the month with the maximum daily precipitation amount is January with 155.8 mm while the smallest occurs in July with 25.3 mm. The most humid month of the year is December with a relative humidity of 76.92% while the driest is July with a relative humidity of 60.64%.

2.2. Methodology

The methodology used consisted of data collection (statistics, land register), creation of maps in Geographic Information Systems (GIS), comparison of results, provision of land use for the year 2020 and finally the development of scenarios.

More specifically, the aim of this chapter was to examine the evolution of land use on the Island of Skiathos on the dates 1945, 1996, 2007 and the current situation in terms of population, infrastructure and development.

With the contribution of Geographic Information Systems (GIS) and the orthophotomaps, the spatial planning of the land uses can be evaluated for all these years and the total area can also be calculated.

Once the thematic maps were created, the model of Cellular Automata (CA) was implemented to make provision on land use development in 2020. In our forecast, human factors such as forest fires were not taken into consideration.

The procedure followed in the study of land use change in the Island of Skiathos was the following:

1. Data input
2. Analysis of land use change
3. Model development
4. Simulation
5. Provision

1. Data input

   The data entered were: land use in 1945 and 1996, distance from roads, distance from the coastline and distance from the main road. Due to the high correlation between the distance from the main road and the distance from the coastline, the distance from the main road was not involved in the model [23,24].

2. Analysis of land use change

   The land use changes were determined by comparing two maps, one of 1945 and one of 1996 [12,16–18].

3. Model development

We used artificial neural networks to create the dynamic transition maps which would be introduced in the model of cellular automata [22].

Five parameters were used in neural networks:

- Neighborhood: size 2 (i.e., 5 × 5 = 25 cells)
- Learning rate, momentum and max iterations number (0.1, 0.05, 1000). These parameters define the neural network training process. Large learning rate and momentum lead to rapid learning. Small learning rate and momentum provide slow but more stable learning. The stability has to do with the large variations in the graph.
- Hidden levels: we defined one hidden layer with 10 nodes—neurons.

4. Simulation

The simulation was achieved by cellular automata. The cellular automata took account of the original map, the factors affecting land use changes and the model developed in the previous paragraph. The cellular automata operates as follows:

- The simulator takes the transition probabilities and calculates the number of cells that have changed.
- The simulator calls the cellular automata model and adds the original map of land use and variables.
- The model scans the cells and calculates the transition probabilities in each class.
- The simulator creates a mesh level "certaincy": each cell defines the difference between the two largest probabilities of lattice transition levels.
- The simulator creates a mesh with the most possible transitions: the cells are transition classes with the highest probability of transition. This mesh level is the auxiliary level to the next step.
- For each class of transition, the simulator is looking to mesh with the most potential transitions the number of cells with the greatest change.

Following the above procedure, the cellular automata gives the result of the simulation for a repetition (iteration). Applying a second iteration, the result of the simulation will be used as an initial land use map. Therefore, in each iteration, we used the previous simulation as initial state land use [1–5,21].

5. Provision

Using the model developed above, we created a forecast for 2020. The prediction was performed using the structure of the above model in which to investigate land use change maps were used in 1945 and 2007 [7,8].

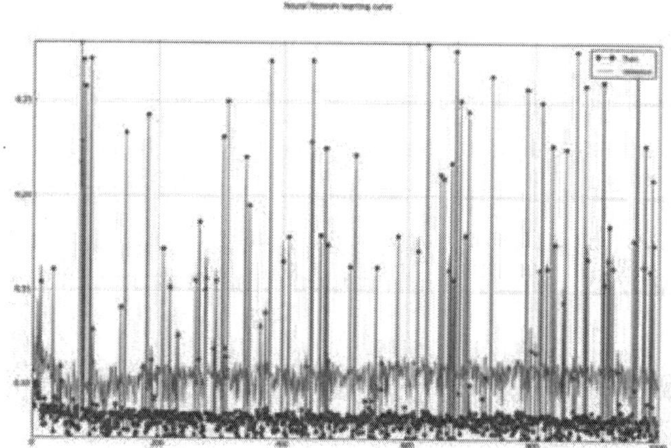

Figure 1. Neural network learning curve.

3. Results and discussion

For the year 1945, we have digitized the land uses in the Island of Skiathos. In Figure **1**, the thematic map resulting from the digitization is presented.

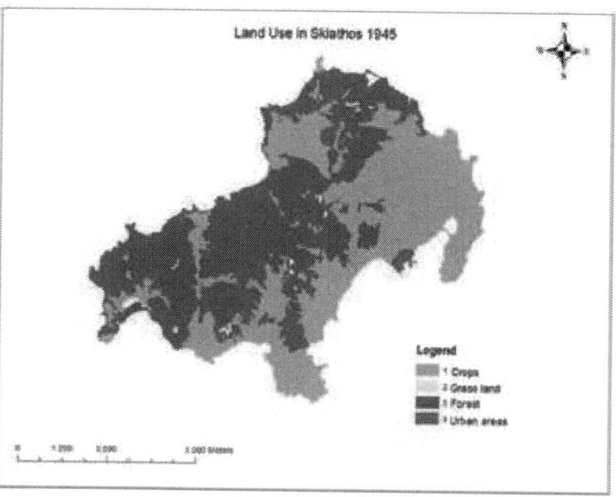

Figure 2. Land uses in Skiathos Island in the year 1945.

In Figure **2**, the digitization and the categorization of the land uses in the year 1945 can be observed. Also, the percentage contribution of each land use type is presented in the Figure (3) and Table (2).

Land Uses	Total Area (ha)
Crops	2229.81
Grass land	28.54
Forest	2413.64
Urban areas	24.29

Table 2. Land uses (total area: ha) in the year 1945.

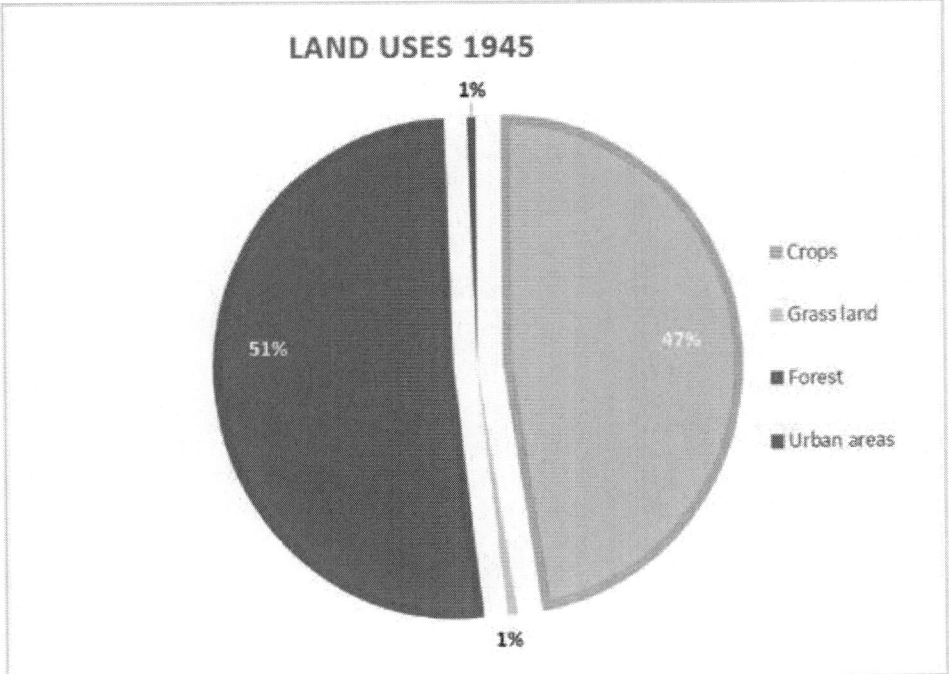

Figure 3. Land uses (% area) in Skiathos Island in the year 1945.

In Figure **3**, it can be observed that the land use with the largest area is the forest (51%) followed by the crops (47%) with a small difference. Finally, the categories grass land (1%) and urban areas (1%) have occupied the same percentage of land and also in very low levels compared to the other two categories.

In Figure **4**, a thematic map of the year 1996 for the Island of Skiathos is presented.

The growth of urban areas is evident from the above map. This means the reduction of the rest of the categories of land uses with the exception of the grass land. More specifically, the percentage of each category is presented in the next chart.

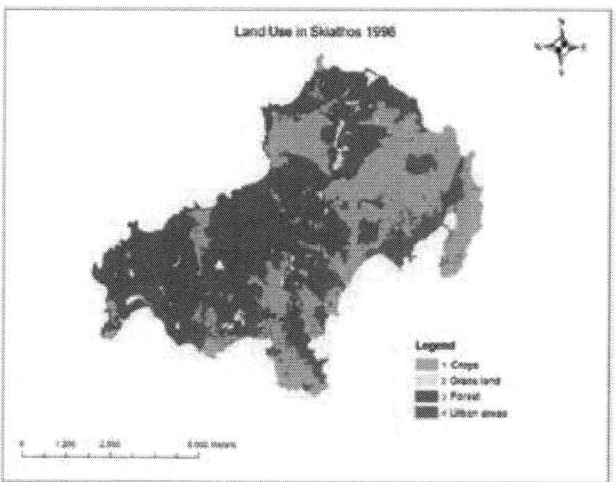

Figure 4. Land uses in Skiathos Island in the year 1996.

Land Uses	Total area (ha)
Crops	1697.57
Grass land	62.82
Forest	2398.26
Urban areas	532.93

Table 3. Land uses (total area: ha) in the year 1996.

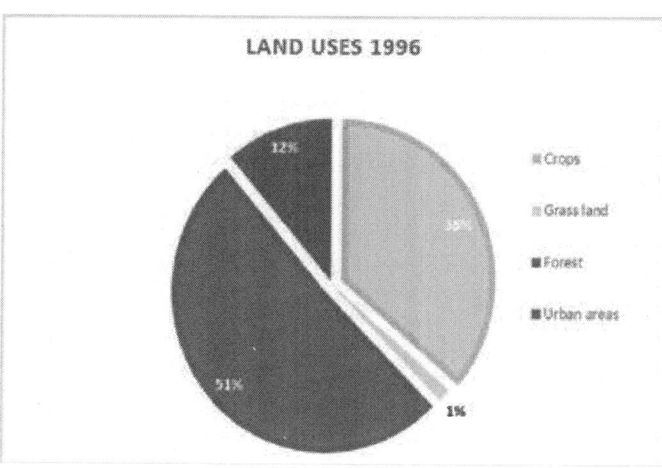

Figure 5. Land uses (% area) in Skiathos Island in the year 1996.

As noticed in Figure **5** and **Table 3** in the year 1996 urban areas have increased (12%). On the other hand, crops have been reduced (36%). Finally, the percentage area of the forests (51%) and the grass land (1%) remained the same.

In Figure **6**, a thematic map of the year 2007 for the Island of Skiathos is presented.

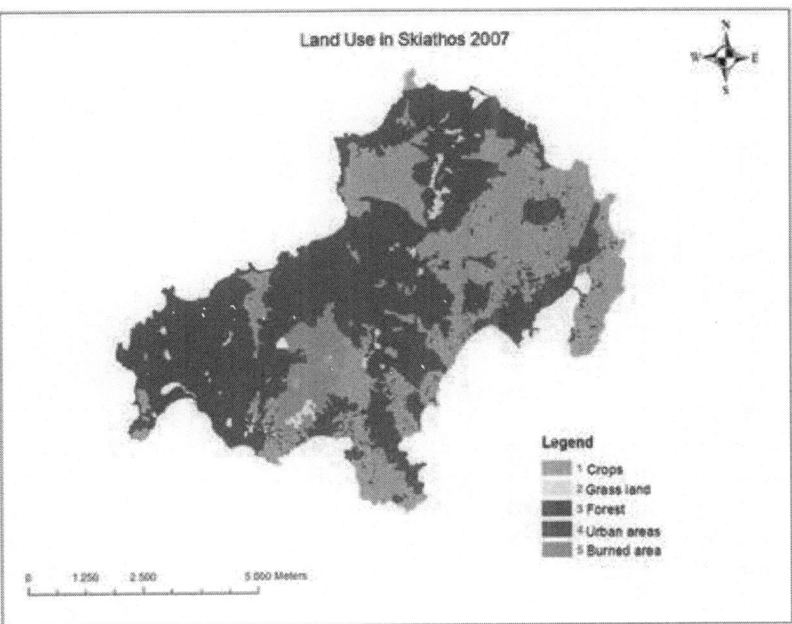

Figure 6. Land uses in Skiathos Island in the year 2007..

In the year 2007, a burned area can be noticed, from the above thematic map. It resulted from a fire in the Island of Skiathos in 2006. The area burnt was previously a forest area. This means that the percentage of forest area has been reduced. This can be observed in Figure **7** and **Table 4**.

Land Uses	Total Area (ha)
Crops	1675.46
Grass land	66.07
Forest	2169.35
Urban areas	553.87
Burned area	219.83

Table 4. Land uses (total area: ha) in the year 2007.

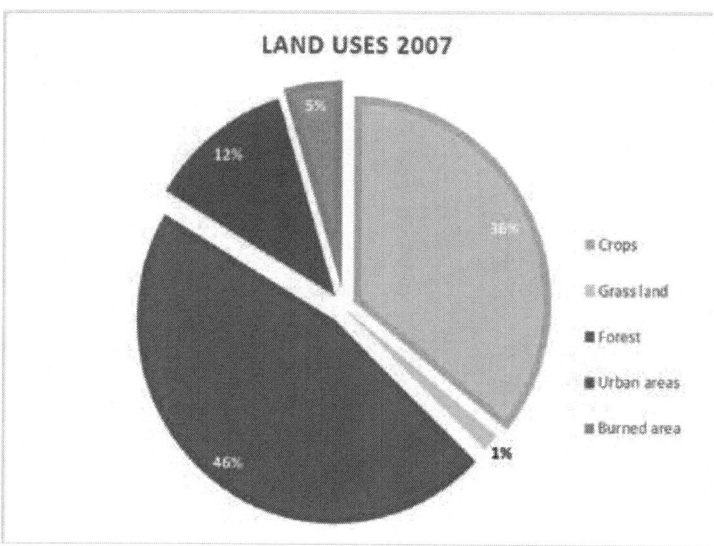

Figure 7. Land uses (% area) in Skiathos Island in the year 2007.

In the year 2007, there was a decrease in the percentage of forest land, while rates of other types of land use remained stable. This is a direct result of the burnt area. From 1945 to 1996, there was a great variation in land use changes in relation to 1996 and 2007. This happened because the interval between the two time periods (1945–1996) is too long (51 years) as opposed to the period 1996–2007 which is only 11 years. That is because there were no data for intermediate dates.

Then, the land uses were significantly changed in these three dates. More specifically, in Figure **8**, the area for each land use can be observed for each year.

In the above chart, a large increase in urban areas in 1996 and the reduction of forest areas in 2007 can be observed. The reduction of the forest areas was mainly due to the appearance of the burnt areas. Also, the crop areas have shown a significant decrease in 1996 mainly due to the growth of the urban areas.

Land Uses	1945–1996	1996–2007
Crops	0.313530517	0.013196376
Grass land	0.545686087	0.049190253
Forest	0.006412983	0.105520087
Urban areas	0.954421781	0.037806706
Burned areas		1

Table 5. Comparison of land uses.

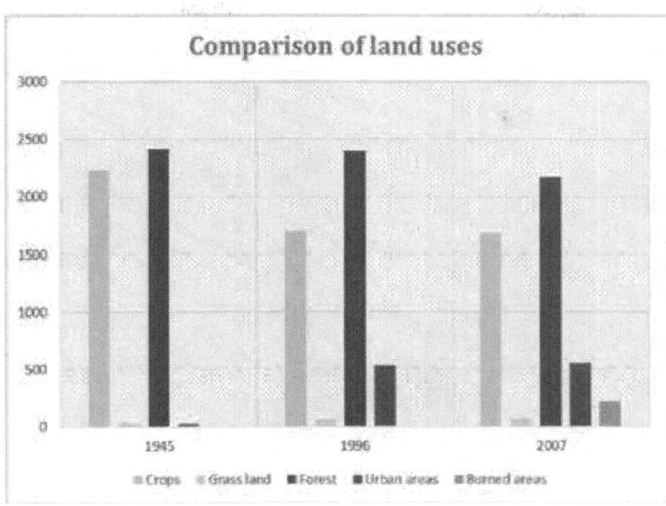

Figure 8. Comparison of land uses.

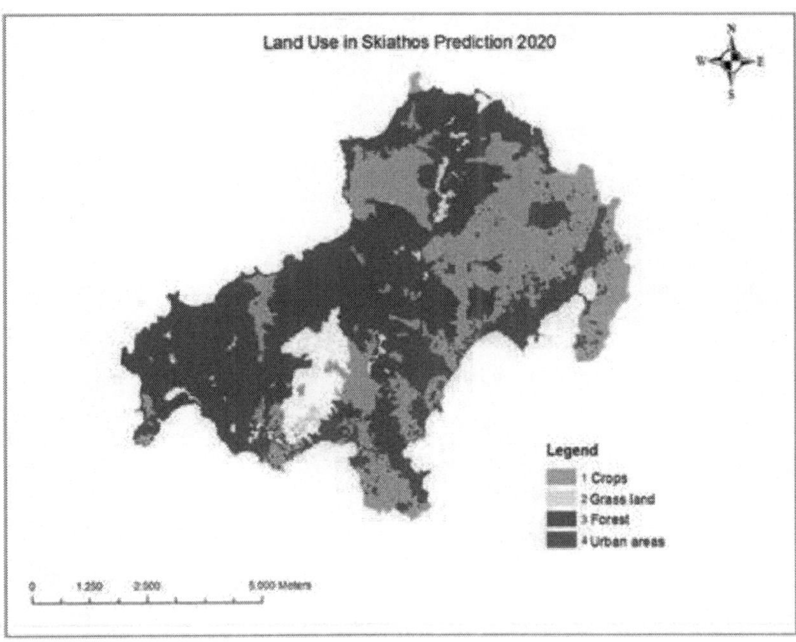

Figure 9. Land uses in Skiathos Island in the year 2020.

Urban areas, between the years 1945 and 1996, largely increased in a percentage of 95% (**Table 5** and **Figure 8**) a thematic map is presented, deriving from the model cellular automata which predicted the land uses in the Island of Skiathos in the year 2020.

As it can be observed in the above thematic map, in the model the burned area was not taken into account, it was left to reforest naturally. In Figure **10**, the percentage of each category of land use is presented, according to the model of cellular automata.

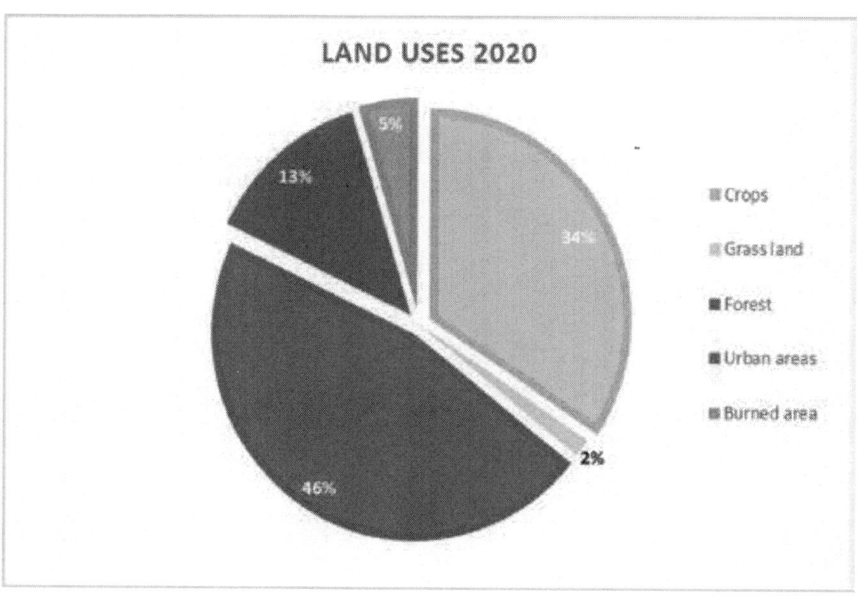

Figure 10. Land uses (% area) in Skiathos Island in the year 2020.

All the categories should be reduced except from the urban area and grass land which should increase.

In **Table 5**, a comparison between the land uses and the years is presented.

Land Uses	1945	1996	2007	2020
Crops	2229.81	1697.57	1675.46	1601.49
Grass land	28.54	62.82	66.07	64.82
Forest	2413.64	2398.26	2169.35	2157.18
Urban areas	24.29	532.93	553.87	618.34
Burned areas	-	-	219.83	219.83

Table 5. Land use (area: ha) in the years 1945, 1996, 2007 and 2020.

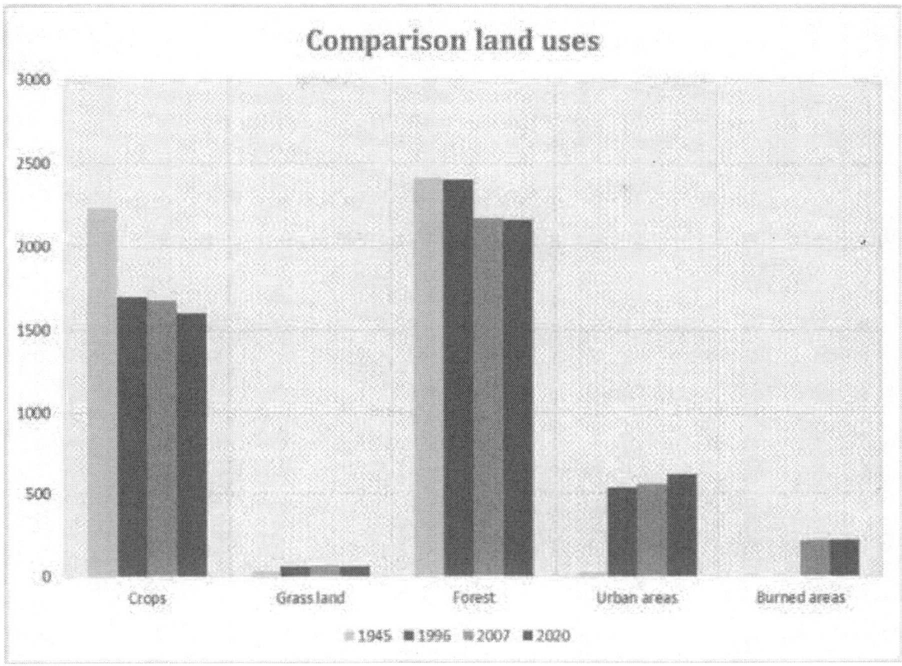

Figure 11. Comparison of land uses in the years 1945, 1996, 2007 and 2020.

In Figure **11**, a significant reduction in the crop areas can be observed between the years 1945 and 1996. In the next years, the reduction was smaller. As for the grass land, there was an increase between the years 1945 and 1996 and a minimum increase in the next years. Also, from the chart it can be observed that grassland occupies the smallest area in the Island of Skiathos.

Forest areas was the category of land use occupying the largest area in the island. However, the greatest reduction in forest areas happened between the years 1996 and 2007, due to the fire which took place in the island.

Finally, the category presenting the greatest increase was urban areas and, more specifically, between the years 1945 and 1996. There was an increase in urban areas during the next periods as well. Burned areas have been recorded in the year 2007 due to the forest fire of 2006 in the south-west part of the Island of Skiathos.

4. Scenarios

Taking into consideration the above analysis and results, two sustainable spatial development scenarios for the Island of Skiathos were proposed. The first scenario is derived from the results obtained from the prediction (application of the model of cellular automata), while the second scenario refers to a more realistic model of development with a focus on environmental protection and sustainable development.

4.1. Trend scenario

This scenario concerns the development of the island in the next years. It results from the application of the model of cellular automata for the year 2020 and shows us an increase in tourism and urban areas.

In the south part of the island, pockets should be created between the urban areas. This scenario does not favor the environment. Also, the sector of the agriculture constantly decreases as the area of gross land reduces in 2020.

Moreover, the northern part of the island will be isolated as there will be no development. The infrastructures in this part of the island are few.

4.2. Sustainable scenario

As an alternative approach, this scenario has a central idea of establishing rules and a comprehensive plan for spatial planning. The new development model is described spatially through the following guidelines—interventions:

- The design and implementation of a Local Spatial Plan for the entire Island of Skiathos
- The implementation regulations

More specifically, we proposed a Local Spatial Plan for the entire Island of Skiathos. More specifically, it includes the following areas:

i. Regulations and incentives

ii. Delimitation of settlements

iii. Land use zones

The regulations include both construction limits and permitted land uses in each area of the island (what is allowed and what is forbidden). It also includes incentives, mainly financial.

4.3. Comparison of the scenarios

The sustainable scenario has more advantages than the trend scenario, as it proposed spatial planning and protection of the environment. Also, there is no development in the south part of the island and particularly an increase in tourism.

5. Conclusions

The diversification of land uses is a result of human activities in the Island of Skiathos. More specifically, it depends on the economic situation, the development and the population of the Island. Also, another parameter is the human factor.

From the above analysis regarding land uses over the years, a great difference between categories of land uses was noticed. What should be mentioned is the increase of urban areas at the expense of forest and crops areas. This happened due to an increase in the population in the island and the increased number of tourists arriving in the island during the last summers.

Therefore, two scenarios, for the Island of Skiathos, were proposed. However, the sustainable scenario offered a more balanced development for the Island of Skiathos. Essentially, the sustainable scenario proposed a rural plan for the whole island.

There has been no other research regarding land uses in Skiathos Island in the past, and, also, this is the first digitization of the area. There is no rural plan for the Island of Skiathos, as well.

References

[1] Batty M.P.: Cellular automata and urban form: a primer. Journal of the American Planning Association. 1997; 63: 266–274.

[2] Benyoussef A.,Boccara N.,Chakib H.,Ez-Zahraouy H.: Traffic flow in a 1D cellular automaton model with open boundaries. Chinese Journal of Physics. 2001; 39 5 428–440.

[3] Clarke kC., Hoppen C., Gaydos L.: A self-modifying cellular automaton model model of historical urbanisation in the San Francisco Bay Area. Environment and Planning B. 1997; 24: 247–261.

[4] Clarke K., Gaydos L.: Loose-coupling a cellular automation model and GIS: longterm urban growth prediction for San Francisco and Washington/Baltimore. International Journal of Geographical Information Science. 1998; 12(7): 699–714.

[5] Couclelis H.: From cellular automata to urban models: new principles for model development and implementation. Environment and Planning B. 1997; 24: 165–174.

[6] Lasanta-MartínezT., Vicente-SerranoSM., Cuadrat-PratsJM.: Mountain Mediterranean landscape evolution caused by the abandonment of traditional primary activities: a study of the Spanish Central Pyrenees. Applied Geography. 2007; 25(1): 47–65.

[7] Levy S.: Artificial Life. New York: Vintage Books. 1992.

[8] Liu Y.: Modeling Urban Development with Geographical Information Systems and Cellular Automata. CRC Press Taylor & Francis Group. 2009

[9] Moreira F., RegoFC., FerreiraPG.: Temporal (1958-1995) pattern of Change in a cultural landscape of northwestern Portugal: implications for fire occurrence. Landscape Ecology. 2001; 16(6): 557–567.

[10] Narumalani S., MishraDR., RothwellRG.: Analyzing landscape structural change using image interpretation and spatial pattern metrics. GI Science and Remote Sensing. 2004; 41(1): 25–44.

[11] PausasJG., Llovet J., Rodrigo A., Vallejo R.: Are wildfires a disaster in the Mediterranean basin? A review. International Journal of Wildland Fire. 2007; 17(6): 713–723.

[12] Portugali J., BenensonI.: Artificial planning experience by means of a heuristic cell-space model: simulating international migration in the urban process. Environment and Planning, 1995; A27: 1647–1665.

[13] Reger B., Otte A., Waldhardt R.: Identifying patterns of land-cover change and their physical attributes in a marginal European landscape. Landscape and Urban Planning. 2007; 81(1–2): 104–113.

[14] Serra P., Pons X., Sauri D.: Land-cover and land-use change in a Mediterranean landscape: a spatial analysis of driving forces integrating biophysical and human factors. Applied Geography. 2008; 28(3):189–209.

[15] SchulzJJ., Cayuela L., Echeverria C., Salas J., Rey BenayasJM.: Monitoring land cover change of the dryland forest landscape of Central Chile (1975e2008). Applied Geography. 2010; 30(3): 436–447.

[16] ToblerWR.: Cellular geography. In Philosophy in geography: 1979; ed. GaleS. and OlssonG., 379–386. Dordrecht: D. Reidel Publishing Company.

[17] TorrensPM.: How cellular models of urban systems work. WP-28, Centre for Advanced Spatial Analysis (CASA), University College, London. 2000.

[18] Uljee I., Engelen G., White R.: Ramco Demo Guide. Work document CZM-C 96.08. 1996; The Hague: Coastal Zone Management Centre, National Institute for Coastal and Marine Management.

[19] Viedma O., MorenoJM., Rieiro I.: Interactions between land use/land cover change, forest fires and landscape structure in Sierra de Gredos (Central Spain). Environmental Conservation. 2006; 33(3): 212–222.

[20] VogiatzakisIN., MannionAM., GriffithsGH.: Mediterranean ecosystems: problems and tools for conservation. Progress in Physical Geography. 2006; 30(2): 175–200.

[21] White R., Engelen G.: Cellular automata and fractal urban form: a cellular modelling approach to the evolution of urban land-use patterns. Environment and Planning. 1993; A,25: 175–1199.

[22] Wolfram,S.: Cellular automata as models of complexity. Nature. 1984b; 311: 419–424.

[23] Wu F.: Simulating urban encroachment on rural land with fuzzy-logic-controlled cellular automata in a geographical information system. Journal of Environmental Management. 1998b; 53: 93–308.

[24] YehAGO., Li,X.: A constrained CA model for the simulation and planning of sustainable urban forms by using GIS. Environment and Planning. 2001B; 28: 733–753.

Chapter 5

Identifying Functionality of Peri-Urban Agricultural Systems: A Case Study

Abstract

Some agricultural systems, especially peri-urban agricultural systems, are characterized as agricultural ecosystems that provide goods and services related to leisure and recreation, the process development beneficial to the environment, such as fixing CO_2, the production of healthy and safe food, and the preservation of natural and cultural heritage. Public intervention in agriculture has traditionally been known as a basic economic task performed by the government whose main objective is food security. But now, agricultural policies have been increasingly challenged by civil society demand, such as a new agricultural model with stronger consideration for non-commodity goods and services. The main obstacles for public intervention are, knowing production of goods and services and externality by peri-urban agriculture system, and identifying what specific demands agriculture needs to satisfy social preferences for goods and services.

We created a descriptive approach for the functionality of peri-urban agricultural systems based on a scientific literature review, which focused on multifunctionality and the goods and services of agricultural systems. This review shows a wide variety of functions that can be grouped according to their economic dimension, social dimension and environmental dimension. We propose a methodology to quantify the functionality of peri-urban agricultural systems by means of indicators.

The proposed methodology was applied to a peri-urban area in Valencia. This is identified as Huerta de Valencia, an agricultural system around Valencia City with a wide variety of resources. The "Comunidad Autónoma de Valencia" government and local government in Valencia City are interested to define a protection scheme and there is open political-institutional debate. The results are very interesting and useful to enrich this debate.

Keywords: Urban fringe, Multifunctionality, Public goods, Joint production, Externality, Market failure, Diversification

1. Introduction

To the conventional role of farming activity as a production sector of raw materials, mainly food products, whose objective was to obtain economic benefits from globalized markets, since the post-productivist paradigm, other capacities have been added: social, economic and ecological implication on the surrounding territory where it has been implemented. To a greater or lesser extent, it is acknowledged that, simultaneous to the production of food and raw material, agriculture contributes to the provision of public goods and services that are related with entertainment and recreation, and to developing beneficial processes, such as CO_2 fixation, healthy safe food production, preserving natural and cultural heritage, and respecting and maintaining the natural environment. The consideration of these aspects results in an agricultural model known as multifunctional [1–7].

It has recently begun to attract considerable attention, with discussion on urban agriculture in industrialized countries. Some agricultural systems on the urban fringe are characterized by being highly valuable spaces that maintain a fragile equilibrium between considering agricultural spaces and/or peri-urban areas that are subject to pressure by alternative land uses (industrial, infrastructure, commercial and leisure centers, etc.). However, society demands of them opportunities to undertake recreational activities, produce quality organic products locally, conserve and maintain spaces relatively near cities and towns under suitable environmental conditions for living, conserving cultural ethnological heritage, etc. [2, 3, 6, 8, 9]. These functions performed by agricultural systems, and the goods and services they can provide, result from the agriculture practice itself, and are generated simultaneously with the production of foodstuffs and raw materials [8, 10–12]. They possess the character of Positive Externality[1] and Public Goods[2],[3] without markets where farmers can interchange them for income, and the existence of market failure that provides theoretical arguments for justifying possible forms of public intervention for correcting these is acknowledged.

Many public goods associated in some form or another with agricultural activities are local [2, 9, 13]. Developing intervention policies requires studying the functional analyses of peri-urban agricultural systems which, beyond the sectorial perspective of agriculture as an economic sector, incorporate detailed analyses of enviro-friendly-related functions, healthy food production, provision of entertainment spaces, etc., as well as the Non Marketable Goods and Services (NMGSs) that they can provide. This work adopts a descriptive approach that takes the multifunctionality of agriculture as a basis to later carry out empirical analyses of peri-urban agricultural systems, the functions they can perform, and the goods and services they can provide.

---

[1] Refers to situations in which the producer of a particular output is not remunerated for it (positive externalities) or does not pay for its detrimental effects (negative externalities).

[2] Refers to goods for which use cannot be easily excluded to other beneficiaries (low excludability) and which can be used various times by different beneficiaries without being destroyed (low rivalry).

[3] Environmental and social functions are known as non-productive, or also as non-marketable, and goods and services are often identified as non-marketable (NMGS). The economic literature about multifunctionality talks indistinctly about (positive or negative) externalities, and about public goods (or "bads"), as most externalities are non-exhaustible, of joint consumption and are impossible to exclude, which render them as public goods.

When applying this theoretical framework to analyze the functionality of any peri-urban agricultural system, it is fundamental to improve the exactness of the spatial scales applied to these analyses by limiting the system under study and suitably detailing the beneficial processes that can be developed.

The Huerta de Valencia is a wealthy agricultural system with a wide variety of resources, but one that is also subject to pressures that can condition its development and survival. Its peri-urban nature (since it is located in the metropolitan area of the city of Valencia), has been greatly conditioned by this fact, and also by the great pressure from competition given the possibilities of alternative activities in this space. You can see the area in Figure 1.

Figure 1. Huerta de Valencia.

From all the above information, we put forward some working hypotheses to respond to our research:

- First: "The multifunctionality framework is a suitable approach to study the functionality of agricultural systems, especially when these systems are characterized by constituting highly valuable spaces, and are acknowledged as performing functions that result in the provision of public goods and services".
- Second: "Peri-urban agricultural systems are characterized as agricultural ecosystems that provide goods and services related to leisure and recreation, the process development

beneficial to the environment, such as fixing CO_2, the production of healthy and safe food, and the preservation of natural and cultural heritage".

- Third: "Developing intervention policies requires a study of the functional analyses done of peri-urban agricultural systems which, beyond the agricultural sectorial perspective as an economic sector, incorporates detailed analyses of the functions that are related to respecting the natural environment, production of healthy foods, provision of recreational scenarios, etc., and NMGS".
- Fourth: "There is concern and social interest about the Huerta de Valencia, and its conservation and preservation".

The first part of this study centres on the multifunctionality concept, and its connection with new sustainable agriculture and multifunctionality approaches of lands, as well as peri-urban agriculture, and provides the most interesting corollaries to define the functions and goods and services that peri-urban agricultural systems can supply, which legitimize public intervention for their conservation. Accordingly, the intention is to specify the supply of public goods and services acknowledged to agricultural systems by defining indicators that allow an empirical analysis. Finally, we include the analysis that we are interested in, the peri-urban system of the Huerta de Valencia.

2. Descriptive approach of peri-urban agrosystems functionality

2.1. Multifunctional, sustainable, and peri-urban agricultural system

Agricultural multifunctionality (AMF) is a conceptual framework within which agricultural activity, which goes beyond the merely productive and commercial domain, forms part of a broader conception by attributing to agriculture functions that care for the natural environment and landscape, conservation of cultural heritage, and functions that improve aspects of the social equilibrium [2]. To these we can add the possibility of providing recreational services, which would perform a recreational function. The fact, that these other (non-productive) agricultural functions exist, which generate NMGS, and for which social demand exists in developed countries, is the basis on which the argument about protecting agriculture and its justification lie.

There has been considerable debate about AMF in different domains. In the political domain, the debate for more than one decade on agricultural policy has been about the multifunctionality of agriculture. The assumed case of this multifunctionality being an intrinsic quality of agriculture as a result of joint production has been underlined in positive AMF approaches. OCDE [4, 5] was probably the first statement who, for legitimacy of policy support to agriculture, adopts the neo-classical economic approach with joint of production of commodities and positive externalities, and market failure for public goods (biodiversity, landscape etc.) as key notions. The legal approach, however, considered it to be the result of the plurality of objectives that the agricultural policy considers and is, therefore, a desirable model [2].

Wilson [7] criticizes that, depending on the research field, the term multifunctionality has been widely and confusedly conceptualized, and goes from agricultural economy to geography, and even moves on to rural sociology. It distinguishes between the MFA concept [11], with its more economic-type discourse, and multifunctionality from a broader landscape and ecology perspective [14, 15], with discourses based on broader and more holistic interpretations. For the first concepts, we could state that multifunctionality is observed from a sectorial perspective, while the second concept seeks the agricultural-rural areas interrelation, and multifunctionality not only of agricultural activity, but also of lands. They point at important implications for analytical research tools such as models for agronomic and agro-ecological relations at farm and regional levels.

Jongeneel et al. [8] defend that multifunctionality in agriculture is directly, although not exclusively, linked to the different functions agricultural land can fulfil. The functions that agricultural land can fulfil include traditional production functions (food, feed), an ecological function (habitat for wildlife), a cultural function (typical landscapes), and a recreation function (enjoying landscape, on-farm attractions and accommodation). Multifunctional agriculture can simultaneously fulfil different combinations of these functions. The goods produced by these functions may be marketable (food, raw materials, ornamental plants, etc.) or non-marketable (wildlife and landscape) [8].

After indicating that the most important non-agricultural functions that it contributes are to help maintain rural areas and the natural environment (in practice this contribution is not easy to measure and is ever-changing). Compés and García Alvarez-Coque [16] defend an agricultural policy that necessarily contemplates reducing negative impacts and agriculture's contribution to mitigating greenhouse effect gases. In parallel, conditions are increasingly imposed and society places more value on positive environmental functions of agriculture, which include CO_2 fixation or maintaining biodiversity, two important cases, and recommend instruments that award surfaces, extensification, quality, sustainable development, biodiversity and rural landscapes.

Renting [2] promotes AMF as a concept and agricultural development course trajectory from an integrated perspective within the sustainable development framework of sustainable development. Reig [12] explores possible connections between multifunctionality and sustainability by understanding that the multifunctional orientation of agriculture leads to measures being taken to correct market failures and to provide coverage of specific environmental functions (protecting biodiversity, landscaping, etc.), which can help enhance social well-being and reinforce environmental sustainability. All this contributes to develop more sustainable agricultural activity.

Chiara [17] indicates that the most significant components of the non- built-up peri-urban areas in some Italian urban areas, are agricultural lands, which usually fulfil environmental, productive, recreational, water balance, and cultural functions, just as multifunctional agriculture does [2, 18]. These multifunctional activities of peri-urban agriculture enrich both the rural and the urban context [19, 20].

Pascucci [21] defends that the multifunctionality of peri-urban agriculture is performed or materializes in the same way as the rural one, insofar as it is the joint production of activity that is firstly oriented to the direct production of market goods and services (food and raw material), but secondly produces goods and services that help land sustainability, preserve biodiversity, and maintain the economic and social vitality of the urban areas in which it is practiced.

Finally, Zasada [22] finds that MFA has been commonly recognized in peri-urban areas. The multifunctional development paradigm provides an approach that strengthens and modernizes peri-urban agriculture. There is a reasonable demand among the urban public for multiple functions and values from farming.

2.2. Provision of goods and services by peri-urban agriculture

Ecosystem services have often been narrowly defined and restricted to natural ecosystems. In recent years, the scope of the ecosystem services concept has been extended, and more studies refer to ecosystem services in agricultural systems. Serious discussion about AMF has resulted from political discussion.

The extensive academic literature available shows recurring concern for the conceptualization of multifunctionality, discussion about the concept, and the supply and demand of the non-commercial functions attributed to it. However, they do not represent an alternative set of compact well-defined contributions, with new contributions on multifunctionality [23]. Nor has a clear definition of the objectives and functions of agriculture beyond the generic (economic, social, and environmental) ones initially established by the Commission been completed.

We reviewed papers that address goods and services of agricultural systems, and that study the peri-urban fringe, its particular characteristics, and especially goods and services related to agriculture. Based on our scientific literature review, we adopted a descriptive approach of the functionality of peri-urban agro systems.

The makeup of peri-urban areas may differ depending on each urban area, the characteristics of the city's growth and systematic land occupancy for town planning, and also on interrelations between town planning and the surrounding rural environment.[4] There are three ways of considering agriculture: land, the living environment and production. Agricultural activity plays a key role in the conception of agricultural peri-urban spaces. In developing countries,[5] agriculture plays an important role in enhancing urban food security and nutrition, local economic development, poverty alleviation, and the social inclusion of disadvantaged groups and sustainable environmental management in cities. In industrialized countries, beyond the

[4] Consult the document, Agricultural management in peri-urban areas [45], which describes the debate one different future models or scenarios for peri-urban areas in France and Italy, according to the interrelation and urban growth over the peri-urban/rural areas near cities.

[5] In November 2001, the following document was presented: Annotated Bibliography on Urban Agriculture, prepared for the Swedish International Development Agency (Sida) by ETC-Urban Agriculture Programme, in cooperation with TUAN and other organizations.

urban sprawl and unplanned situations, agriculture is taken into account to: care for space (fertile land is a non-renewable resource); provide a view of agriculture in land tenure terms (agriculture is an economy activity and land is an economic resource); contribute to urban employment and to reduce inequalities; favour the conservation of memory and the roots of agriculture (it forms part of the territory's identity); and contribute, through its environmental functions, to management of water and natural risks, to fight against greenhouse effects, to value waste, and to increase the value of landscapes and tourism. This peri-urban scenario is enhanced by AMF.

It is not unreasonable to indistinctly talk about the agricultural areas located in urban areas where agricultural activity is a multifunctional economic activity, and about peri-urban areas of disperse urban growth where agricultural activity is actually marginal; that is, in a small space in local interstitial areas, or in an almost continuous urban matrix, but one that is broken by small closed agricultural spaces. Isolated agricultural areas influence the main agricultural functions, and hence agri-urban development. This isolation is a disadvantage for small fields given conflicts with local inhabitants, high land prices and low income. Non-agricultural work opportunities are scarce, and the share of part-time agriculture and hobby farming is large. These areas can be termed infra-urban.

The interest of the present work lies in peri-urban systems, where the productive structure and agricultural areas are almost continuous, although there is a separation between the urban and the agricultural/rural matrix. In these agricultural areas, agricultural land is more available, and the productive, environmental and landscape functions of agriculture are more important.

In relation to agriculture, and its multifunctional activity consideration, Reig [24], considers that it produces a wide range of goods, and includes a diagram to summarize and classify them as public and private. Private ones include the production of food and raw material of an agricultural origin, rural tourism, and other possible private goods. Public goods are classified as: 1) environmental: protecting landscape values, protecting biodiversity and protecting land, plus controlling erosion; 2) social: contributing to the feasibility of unfavoured rural areas, protecting cultural values in rural areas, and protecting rural areas from population loss.

Specifically for peri-urban agriculture, Zasada [22] reviews societal demands and the provision of goods and services by farming. Beyond traditional agricultural functions, peri-urban agriculture is increasingly acknowledged for its deliverance of local food, as well as recreational, educational and other social services: Environmental and landscape amenities, Leisure and recreation, Regional food supply which directly, Lifestyle farming, Recreation-oriented diversification, Social farming, Short supply chains and Direct marketing.

If we consider one of the broadest and more specific definitions for sustainable agriculture, that by Gómez-Villarino et al. [25] is in accordance with the basic functions that the natural environment performs as a source of resources, a receiver of effluents and a support of activities, sustainable agriculture will be orientated to the rational exploitation of renewable resources, depending on their renovation rates, will not emit effluents to nature that exceed its assimilation capacity, and will respect ecosystems' hosting capacity when inducing transformations in them: it will promote savings and efficiency in water resources, use of alternative irrigation sources (wastewater), energy savings and efficiency, use of renewable energies, land management that avoids erosion phenomena, integrated and ecological production methods; it will avoid practices that pollute water, will respect the natural vocation of ecosystems, and will limit accessibility to the whole land; it will conserve traditional buildings and constructions, the makeup of plots and farms, and separation elements: walls, edgings, hedges, dispersed trees. It will also defend these functions: territorial equilibrium: fixing a dispersed population in the land to counteract the concentration tendency; creating a social system for activity by developing population settlements and constructed elements to also confer a material value, and historical and cultural interest; it will be a support for infrastructures and more productive economic activities, whose sustainability lies in respecting the natural vocation of various types of ecosystems.

The more an agricultural system approaches the pattern defined by Gómez-Villarino, the further it will move from productivity patterns.

The research by Ivesa and Kendalb [26] lists the following as components of the value of peri-urban agricultural areas: Culture (agricultural land near cities is an important part of cultural heritage), Education (farms near cities are important places for people to learn about the natural world, Agricultural land near cities provides good opportunities to teach people about where their food comes from), Environment (farmland near cities is important because it protects natural landscapes. The agricultural land closest to where people live is valuable because it contains natural areas not found in the city. Land near cities used for farming is not a place with worries about protecting native species of plants and wildlife. I like to know that native plants and animals are able to survive on the edge of cities), Aesthetics (people enjoy the scenery of agricultural landscapes on the edge of cities. It is important that people can see farm animals near where they live), Food security (farms near cities should produce food to make sure that people who live there have a secure food supply), Food availability (it is important that people have access to food produced near where they live. We can still have a fresh supply of food without preserving agricultural land near cities. People prefer to buy locally produced food).

As a result of studying and analyzing these works, we proposed a descriptive approach of the multifunctionality of peri-urban agricultural systems, and the goods and services they can provide, by differentiating private ones from public ones, and among these, positive and negative externalities. We classified the various functions depending on whether they are economic, social or environmental.

We considered the following functions and/or goods and services:

- Configuring natural heritage: Agricultural systems are local in nature, and the surrounding conditions they develop in (climate, soil type and characteristics, edaphic properties, availability of water resources, the land's orography, etc.), condition the configuration of true natural heritage in accordance with crops, the production practices and techniques applied, and the agricultural structures that are implemented.

- Creating new landscape forms: as a result of the previous point, the generated landscape is unique and irreplaceable, and is, more often than not, high quality. The impact of agricultural activity is attractive landscapes, which is why citizens are very sensitive about their conservation and maintenance.
- Conserving biodiversity: the diversity of ecosystems and the ecological processes developed by agricultural practice has led to much heritage being accumulated as biological diversity and ways of life.
- Protecting water resources: in quality terms, distribution in time and quantity for urban, rural, industrial and hydroelectric use by protecting and sustainably using aquifers, springs, and water sources in general, protecting and recovering basins and microbasins, etc., and introducing practices that help minimize and correct negative externalities that generate diffuse pollution.
- Producing energy: agriculture is a source of biological material and products to generate biomass.
- Mitigating greenhouse effect gases by fixing, reducing and storing carbon (CO_2) and other greenhouse effect gases.
- Scenic beauty: it results from the presence of forests, natural landscapes and biodiversity elements that are appealing and act as a basis to develop various forms of tourism: ecotourism, sun-and-beach tourism, scientific, observation and adventure tourism.
- Maintaining areas such as forests, wetlands, reefs and mangrove swamps to mitigate impacts of disasters caused by flooding, landslides, drought, etc., which are associated with natural phenomena.
- Protecting soil: agricultural activity plays a key role in soil formation and soil conservation processes as it favours organic matter accumulation and soil fertilization by nutrient fixation.

However, agricultural activity can also have negative effects, or negative externalities, such as:

- Consuming water resources: Irrigation agriculture consumes vast amounts of water resources, which restricts their availability to other sectors and ecosystems. In some cases, aquifer overexploitation can have very negative effects on nearby ecosystems, which can even disappear. As for quality of water resources, irrigation agriculture, but also intensive farming, has caused major problems by contaminating aquifers because growing concentrations of leached nitrates accumulate. Thus, practices need to be introduced that help minimize these effects, implement more efficient irrigation technologies, and help minimize and correct the negative externalities that diffuse pollution generates.
- Healthy safe food: Irrigation agriculture, especially horticulture, intensively uses phytosanitary products: fungicides, herbicides, insecticides and phytoregulators. Its use of others types is more restricted, such as nematicides, acaricides, etc. They can spell enormous problems from the presence of waste in food, persistence and accumulation problems in soil

which, through leaching, can affect bodies of water. As regulations progressively tighten the use of these substances, they are used less. The inclusion of good agricultural and livestock practices, and an agriculture that respects ecological and integrated production systems, are not only necessary to comply with laws, but to also meet new consumer food trends; and to attend consumers who are concerned about the health of food, traditional production with traditional varieties and cultivars that offer excellent taste quality, production by applying traditional more enviro-friendly working techniques, and production obtained from family-run farms related with traditional ways of life in a nearby agricultural space that offers good quality of life.

Regarding the use and enjoyment of peri-urban agricultural spaces as recreational and entertainment spaces, outsourcing activities is feasible in such a way that some surfaces would no longer remain agricultural, but would be occupied for other uses, termed land activities which, by adding value, sum complementary income to the primary value. Such activities include rural tourism as various ecotourism forms, sun-and-beach tourism, scientific, observation and adventure tourism: e.g., birdwatching and wildlife, hiking, cycling, fishing, hunting, and even swimming. Accordingly, the positive effects set out in the previous section that agricultural activity has on land also apply to such spaces. In these cases, other production activities should also maintain an environmental function that respects nature and biodiversity, as well as other factors, such as the landscape and cultural heritage, on which the agricultural system's multifunctional nature is sustained.

		GENERIC SERVICES	Functions	Goods and services	E	S	M
PRIVATE		0. Land	Heritage	Land as heritage	X		
		1. Food production	Primary production of foodstuffs	Provides vegetable foods, seeds, plants, and edible fruits, and biological material	X		
		2. Recreation	Generator of tourism	Provides space for eco-tourism, fishing, hunting, and relaxing.	X		
		3. Energy production	Biomass	Provides biomass (plant remains and energy crops).	X		
PUBLIC	E. POSITIVES	4. Configuration of natural heritage and creation of new landscapes	Configuration of a farming system that contributes to an agrarian system by creating an agricultural landscape of great value	Natural heritage and beautiful landscape			X
		5. Production and conservation of bio-diversity	Environment for animals and plants	Enables development of biological flows. Enables preservation of plant resources: populations of important species such as pollinators, native species, rare or threatened species			X
		6. Protection of water resources	Water storage	Improves water supply			X
			Frees resources by improving irrigation efficiency				X
			Frees irrigation resources with alternative sources				X
		7. Mitigation of greenhouse effect	Sink for greenhouse gases	Regulates climate by regulating greenhouse gases			X
			Reduction of emissions	Reduces greenhouse gas emissions using alternative energy sources			X
			Regulation on atmosphere chemical composition	Contributes to improving air quality			X
		8. Mitigation of disasters caused by floods, landslides, and droughts	Regulation of water flows	Improves ability to respond and adapt to natural disasters: storms, floods, droughts			X
			Retention of sediments and erosion control				X
		9. Territorial and cultural scientific	Territorial equilibrium	Articulates nuclei and avoids congestion		X	
				Creates jobs		X	
				Makes business more dynamic		X	
			Cultural use	Provides a unique cultural heritage		X	
				Provides space for education		X	
			Scientific use	Provides important scientific elements: practice, vocabulary, tools, infrastructure		X	
		10. Land protection	Soil forming process	Favours soil formation.			X
			Soil conservation	Favours accumulation of organic matter and soil fertility by fixing nutrients			X
	E.NEGATIVAS	11. Water resource protection	Diffuse pollution	Causes loss of quality of groundwater through pollution of aquifers			X
			Degradation of wetlands	Damages surrounding wetlands			X
		12. Production of safe healthy food	Risks to human health from pesticide residues	Interferes with supply of healthy and safe food		X	

Table 1. Peri-urban agricultural system: functions, goods and services.

The attached **Table 1** includes a list of the goods and services that an agricultural system can provide, distinguishes between those considered private from public, and those with a positive or a negative externality nature. All goods and services are classified according to whether they are economic, social or environmental. Up to 12 generic services are defined, which are subdivided into 22 functions and 22 goods and services. This is contemplated as a descriptive approach to AMF as a basis to later carry out empirical analyses on peri-urban agricultural systems.

The public goods that agriculture generates differ from some agricultural systems to others depending on different aspects of the environment, agricultural practices, etc. Consequently, the supply and demand of public goods is extremely heterogeneous.

Regarding the joint production characteristic, interdependence exists between agricultural production and production of other goods (NMGS), regardless of their technical or biological nature. Moreover, some production factors exist from which several different outputs are obtained, which cannot be separately assigned to each one. Reig [9] considers that joint production is related with the production techniques applied, environmental conditions, technology and the subsequent allocation of resources, etc., therefore knowledge of it demands multidisciplinary research efforts, and must be oriented to specific geographic locations.

Supply will be more heterogeneous as a result of differences between the physical basis (climate, soil type, quality of irrigation water, etc.) on which the activity in different regions lies, and also due to the agricultural activity management at the exploitation level with intensification or extensification of production systems. The fact that externalities exist and that market failure is acknowledged will trigger intervention. However, knowledge about the joint production process is essential to select and design intervention instruments that can promote the production processes or systems that generate the positive externalities or public goods that are in demand [27].

3. Analysis of the functionality of agricultural systems

Conducting studies about multifunctionality entails defining the significant indicators which suitably explain the underlying processes and relations that link the agricultural activity and functions, and the goods and services that it provides. These indicators should be defined and proposed in different case studies and in accordance with not only the most relevant aspects of the agricultural system to be studied, but also with the purpose of the study in question, but mainly according to existing sources of information, etc. **Table 2** includes the defined indicators.

For a multidimensional assessment, the next task is to identify appropriate indicators of economic, social and environmental dimensions of multifunctionality in peri-urban areas.

Specially, we want select a limited number of attributes which can be quantified, in order to evaluate the performance of agriculture, and determine its multifunctional character, in different settings and situations.

This selection was made by review of literature about some indicators, proposed by some authors related to economic dimension [20, 28–31] to social dimension [32, 33], and to environmental dimension [22, 34, 35]. There are some works which integrate all dimensions together [2, 36, 37].

Were used too, some reference papers, from statements, public institutions, government institutions, etc. which include information about some indicators in different dimensions:

- OCDE 2001: Environmental Indicators for Agriculture Volume 3: Methods and Results [38].
- OCDE 2002: Effects of agricultural policies and practices on the environment: review of empirical work in OCDE countries. COM/AGR/CA/ENV/EPOC(2001)60/FINAL [39].
- Proyecto WADI: The Sustainability of European Irrigated Agriculture under Water Framework Directive and Agenda 2000 [40].
- COM (2006) 508 final, COMUNICACIÓN DE LA COMISIÓN AL CONSEJO Y AL PARLAMENTO EUROPEO. Establecimiento de indicadores agroambientales para el seguimiento de la integración de las consideraciones medioambientales en la política agrícola común.
- OCD 2008: Environmental Performance of Agriculture since 1990: Main Report, Paris, France (2008).

As a result, a set of indicators were selected (**Table 2**), designed to verify and control the multifunctionality of agriculture farming system level. **Table 2** includes indicators used to evaluate the peri-urban agricultural system of Huerta de Valencia.

[illegible]	[illegible]	[illegible]	[illegible]	[illegible]	[illegible]

Table 2. Peri-urban agricultural system: indicators for functions, goods and services.

Indicators need the requirements [38] that they are:

- Policy-relevant: indicators should be related to the government decisions and other stakeholders, and results, in the agriculture sector;
- Scientific and analytically basis: indicators based on science, with possibility to be developed and improved in progress;
- Measurable: indicators from available data, with reasonable collection cost;
- Simple interpretation: indicators should be easy to interpret by policy makers, stakeholders or the wider public.

3.1. Indicators for economic function

Indicators that explain the economic function are related to indicators on economic viability of farming. They are taken into consideration: 1) Farm real estate values. It is the principal source of collateral for farm loans, enabling farmers to finance the agricultural activity. 2) Farm income as indicator of economic performance. Also, rents resulting from possible alternative activities that can develop in the field of agriculture, more related to the possibility of out-sourcing activity and development related to the possibilities of using agricultural areas for leisure and recreation activities. Considering potential income, as complementary income from farming tourism. Calculation of income as defined by the OECD [38] as the difference between the value of gross output and all expenses, including depreciation at the farm level from agricultural activities.

The use of agricultural areas for leisure and recreation is related to rural amenities from landscape, biodiversity, and wildlife habitat, the existence of a cultural heritage related to traditional agricultural practice: architecture, civil engineering, agricultural engineering, hydraulic engineering, etc. In many cases, specific local environmental and external factors drive urban farms to develop unique innovations for space-intensive production systems, often creating a dominant paradigm for urban farming for a given location [28].

3.2. Indicators for environmental function

The environmental function is related to the configuration of farming systems that contribute to the creation of characteristic agricultural landscapes of great value, to the conservation of biodiversity, to the conservation of the soil, to the protection of water resources, etc.

3.2.1. Configuration of Natural Heritage and Creation of new landscape shapes

Agriculture is, in many cases, responsible for shaping landscapes and ecosystem with specific characteristics. Managing human intervention over nature, the crops, cultural practices, plot size, shape of the plots, alignment of the plots, and infrastructures creation, can contribute to the creation of agricultural heritage, natural with great value. These agricultural landscapes deliver cultural and recreational, services, which do not sustain agricultural production, but deliver benefits derived from the aesthetic function of landscapes, including tourism, sense of place, spiritual experiences and recreation, offering possibilities for additional income through, e.g., the recreation and tourism sector [41].

Agricultural landscapes are characterized by their spatial structure and composition. Landscape structure includes the diversity and complexity of the spatial (and temporal) structure of the landscape: orography, slope of the land, spatial organization of fields, plot size, shape of the plots, alignment of the plots, and infrastructures creation. Landscape composition refers to the relative prevalence of land use/land cover types, crops types, and landscape elements: water, trees, etc.).

Indicators proposed are:

- Soil cover as number of days in a year that the soil (agricultural land) is covered with vegetation [38]. See you Table 3
- Mosaic or tessellation is interpreted by agricultural matrix more or less plotted, more or less broken in small agricultural spaces, with a great variety of crops. Mosaic can be interpreted by number of agricultural plots, and variety of crops.

Farmland		**Arable land**	
Farms	Ha	Farms	Ha
10.030	21.975	9.079	20.123
	100%	91%	92%

Table 3. Arable land in Huerta de Valencia.

	Total farmland surface (Ha)		**Arable land (Ha)**	
	Number of farms	**Total surface(Ha)**	**Number of farms**	**Total surface (Ha)**
From 0.1 to 0.2 Ha	5	0.38	3	0.35
From 0.2 to 0.5 Ha	1,161	389.34	1,043	353.37
From 0.5 to 1 Ha	1,283	885.50	1,261	893.84
From 1 to 2 Ha	1,766	2,448.25	1,760	2,424.21
From 2 to 3 Ha	921	2,196.66	892	2,113.11
From 3 to 4 Ha	493	1,668.58	474	1,597.20
From 4 to 5 Ha	298	1,307.19	271	1,188.70
From 5 to 10 Ha	632	4,254.70	604	4,056.80
From 10 to 20 Ha	256	3,415.14	246	3,264.29
From 20 to 30 Ha	67	1,602.28	56	1,348.93
From 30 to 50 Ha	36	1,359.24	28	1,039.07
From 50 to 70 Ha	13	727.76	11.00	621.82

	Total farmland surface (Ha)		Arable land (Ha)	
	Number of farms	Total surface(Ha)	Number of farms	Total surface (Ha)
From 70 to 100 Ha	6	495.34	4.00	322.67
From 100 to 150 Ha	4	510.74	2.00	262.70
From 150 to 200 Ha	1	150.00		
From 200 to 300 Ha	1	200.00	1.00	200.00
From 300 to 500 Ha	1	399.71	1.00	399.71
Total	6,837	21,974.87	6,837.00	20,122.77

Table 4. Farm size in Huerta de Valencia.

There are a lot of very small size plots. **Table 4**, indicates that there are arround 22.000 farms. From these, 60% are short farms, smaller than 10 ha. 25% between 4 and 10 Ha.

Figure 2. Spatial structure and composition.

There are a wide variety of crops (**Table 5**): traditional orchard crops, grain represented by rice, and a lot of citrus.

Vegetables		Other arable crops
ARTICHOKES	CANTALOUPE	Alfalfa
EGGPLANT	CUCUMBER	FLOWERS
ONION	WATERMELON	ORNAMENTALS
LETTUCE	TOMATO	**CITRUS**
CABBAGE AND CABBAGE	GREEN BEAN	SWEET ORANGE
CAULIFLOWER	GREEN BEAN	MANDARINO
ESCAROLE	SPINACH	LIMONERO

Vegetables		Other arable crops
GRAIN CEREALS	**TUBERS HUMAN CONSUMPTION**	**FRUIT**
	Potato TOTAL	GRANADO
RICE (PADDY)	CHUFA	ALGARROBO

Table 5. Variety of crops.

3.2.2. *Biodiversity protection*

Variety of natural conditions and farming traditions can create unique landscapes that are not only pleasing to the eye but provide the living conditions for many plants and animals. Differing farming practices have led to a variety of agricultural habitats that host a large number of plant and animal species. Maintaining adequate farming practices is therefore key to biodiversity conservation. Biodiversity generally decreases when the intensity of farming increases. On the other hand, land use change affect negatively, and abandonment is considered detrimental to biodiversity. So, extensive farming systems are most vulnerable to abandonment, and so to decrease biodiversity. Small farm size usually entails the consolidation of fields boundary, such as hedges and small trees, structures for feeding, nesting and shelter against predators. These are important for the crop protection [2]. Farmland birds are too, indicative of overall biodiversity.

In the area of Huerta, we find sites hosting natural diversity of habitats subject to different forms of protection. South of the Huerta de Valencia is the Albufera Natural Park. Paddy field area is located within the Albufera Park.

Figure 3. Paddy fiels area in Albufera Park.

North of the Huerta de Valencia, is the wetland "Rafalell I Vistabella", and "Marjal del Moro". Both are protected wetlands included in the Ramsar Convention. Albufera is also a wetlands. Tree areas are included in Nature 2000. Ecosystems located into or in the immediate vicinity and bordering on agricultural land.

The structural composition and dynamics of these ecosystems can´t be understood but as a result of interdependencies with the surrounding agricultural areas and certain agricultural activities. In all cases, irrigation returns is very important for wetland. Moreover, the agricultural environment intervenes in biological flows, allowing the species movement through connection elements present, such as trails, enclaves and hedges that include either through called frames biophysical, hydrological flow to the surface. The canals and drainage system, and irrigation canals connected, in all cases, with these spaces, forming a waterways system in some cases with abundant marsh vegetation that act as ecological corridors, linking together different spaces. A lot of species use marsh vegetation in canals for feeding, nesting and shelter.

Indicators proposed:

- Relationship between the surface of protected areas and the area under crops. 14,33%, according to **Table 7**.
- Trends in population distributions and numbers of bird species related to agriculture. We can see the annual inventory of species in protected areas. Regional administration make every year an annual inventory. In these inventories we have found many bird groups (rallidos, ducks, waders, terns, etc.) but also fish fauna among which include eel (*Anguilla anguilla*), mullet (*Mugil cephalus*) and especially the fartet (*Aphanius iberus*), amphibians as samaruc, and vertebrates.

On the other hand, biodiversity that is dependent on agricultural activities and/or affected by it, is related with the diversity of crop varieties/livestock used in agricultural production. These indicators help to reveal the resilience of agricultural production to environmental changes and risks which occur through diversifying the number of varieties/breeds in production.

- "non-native" species threatening agricultural production and agro-ecosystems.
- Crop varieties/livestock breeds that have been registered and certified for marketing.

Natural area protected	Type
Albufera	Parque Natural, LIC, ZEPA, Zona Húmeda
Marjal dels Morors	LIC, ZEPA, Zona Húmeda, incluye microrreserves
Marjal de Rafalell i Vistabella	Zona Húmeda
Parque Natural del Turia	Parque Natural

Table 6. Natural area protected.

Total farmland surface (Ha)		Arable land (Ha)		Natural area protected (Ha)
Number of farms	**Total surface (Ha)**	**Number of farms**	**Total surface (Ha)**	
10,030	21,975	9,079	20,123	3,150
	100%	91%	92%	14.33%

Table 7. Natural protected area for arable land.

3.2.3. Water use

Agriculture has capacity to interfere on the supply and quality of water to improve it. Agriculture performs storage functions and water retention. Agriculture regulates the water regime in the crop area and in bordering wetlands, already named above.

The share of agriculture in total water utilisation is normally high. Irrigation technology help to achieve economic efficiency by management and distribution.

We are developing three indicators related to agriculture's use of water:

- Irrigated system/ Absolute volume of water used per arable area:
  - 1) Irrigation efficiency: m3/Ha
  - 2) Share of irrigation water applied by different forms of irrigation technology (**Table 8**).

Total Surface (Ha)	Irrigation method (Ha)			
	Sprinkler	Drip	Flood	Other method
19,044	253	4,886	13,782	122

Table 8. Surface by irrigation method (Ha).

The least efficient method, flooding, is the most used irrigation method in Huerta de Valencia. This suggests that water be used more efficiently. Water is considered a scarce resource and consequently issues related to irrigation efficiency are of importance. But, the irrigation method, is important for another function:

- The reservoirs and canals provide habitat for waterfowl and fish.
- Wetlands around agricultural areas receive seepage and drainage from canals.
- Irrigation canals provide habitats.
  - 3) Supplying adjacent wetlands: relationship between feedback water resources that supply adjacent wetlands of all the resources used in arable area.

3.2.4. *Climate change and global warming contribution*

It is now widely believed that the increased atmospheric concentration of greenhouse gases (GHGs) is contributing to the process of climate change and global warming.

Huerta de Valencia as agrarian system has capacity to act as a GHG sink, regulation of local and global climate change, and contributes to atmosphere composition regulation, and the improve of quality air. But the cover cropping, tillage systems and crops management in Huerta de Valencia make these capacities difficult, because the most important crops are vegetable and there are few trees.

3.2.5. *Soil protected, land conservation*

Appropriate land use, combined with environmentally sound soil and water management practices can help to reduce the peak flow of surface water and loss of soil sediment.

3.2.6. Negative externalities

Environmental value includes negative impacts of farm management practices. These negative impacts refer to pollution derived from intensification in irrigation crops, and from pesticides applied.

Huerta de Valencia is an agricultural system which practices farm management that can induce environmental damage. Nitrogen (N) and phosphorus (P) surplus in excess can run-off into surface waters and percolate. So it is important to consider the risk of groundwater pollution. Regional administration has rated the area with high risk of groundwater vulnerability.

- Water quality risk indicator: Potential concentration of nitrate (or phosphorus) in the water flowing from a given agricultural area, both percolating water and surface run-off.
- Water quality state indicator: Nitrate concentration in water in agricultural areas when these areas are included as vulnerable areas.

It is important to have analysis from soil and water, to adjust surplus N and P to the crop needs.

Pollution by pesticide, include soil, water and air pollution. The most important result is the potential risk to human health and the environment. It required pest management, to reduce the environmental and health impacts of pesticide use. Recommended farm management practices can reduce pesticide residues left in soil, by lowering the quantity of pesticides applied and using less toxic and less persistent pesticides. Indicator:

- Adoption of IPM by vegetable growers. Land management practices: Share of the total crop area under environmental land management practices.
- Provision of agricultural employment opportunities

3.3. Indicators for social function

Social goods and services of Huerta de Valencia derive from territorial and cultural-scientific elements.

3.3.1. Equilibrium territorial

Urban expansion has been the most important driver of very high consumption of land and agricultural resources. The urban fringe is often characterized by uncontrolled urban development that results in discontinuous patterns and consequent fragmentation of farmlands. As a result, it becomes a particular contemporary peri-urban landscape, where residential low-density settlements are intertwined with farmlands that have been partially modified and reduced by urbanization. This is the main landmark of metropolitan areas [42].

In Mediterranean cities, for decades, small residential urban settlements around the big city, absorb some of the population growth (Barcelona, Valencia, Milan, etc.), creating major metropolitan areas very densely populated. Agriculture involves this urban development process with opportunity of working, building communities, connect people to each other and the land, family cohesion, producing food and other products for consumption and for sale.

Actually, in these developed cities, agriculture, rather than food production per se, takes on an important role in providing recreational opportunities for citizens (recreational routes, food buying and meals on the farm, visiting facilities) or having educational functions (bringing youth in contact with animals, teaching about ecology, etc.).

Huerta de Valencia has been shaped historically by the farmers, from small settlements, the network of canals and historic trails, and scattered farmhouses and "barracas"[6]. As a result of agricultural activity, it has developed important agronomic heritage (agricultural practices, varieties, tools, vocabulary, etc.), hydraulic heritage necessary to water resources management, and civil architecture (mills, farmhouse, etc.). All together, these make Huerta de Valencia an agricultural system with great cultural and scientific value.

In Valencia, from 1950, economic growth resulted in very important changes in the city and pre-metropolitan settlements. Population in this area has exhibited a considerable growth. In small settlements, this is faster than Valencia City. You can see Table 9.

Year	Now VMA	Valencia	(VMA/Valencia)x100
1877*	88,000	165,000	53.33
1900*	110,000	209,000	52.63
1950*	214,000	450,000	47.56
1975*	470,000	718,000	65.46
2008**	758,232	771,466	98.28
2011	746,696	798,033	93.57

Table 9. Population evolution in Valencia Metropolitan Area (VMA).

On the other hand, agriculture provides agricultural employment opportunities. De Zeeuw (2003) noted that in addition to the economic benefits to the urban agricultural producers, urban agriculture stimulates the development of related enterprises: the production of necessary agricultural inputs and, the processing, packaging and marketing of outputs. The activities or services rendered by these enterprises may owe their existence wholly or in part to urban agriculture [43].

Indicators:

- Employment: agricultural employment in total civilian employment.

Agricultural sector has a declining role in overall economic activity. This is reflected in the decrease of farm employment.

[6] Typical farmhouse in Huerta de Valencia.

	C. Valenciana			Spain		
	Total	Men	Women	Total	Men	Women
farming	4.2	5.5	2.5	3.3	4.9	1.3
industry	14.7	19.7	8.3	16.7	22.5	9.3
construction	10.0	16.4	1.8	10.8	17.9	1.5
services	71.1	58.5	87.5	69.2	54.7	87.9

Table 10. Employment by economic sector, 2009.

If we consider employments for the last years, by activity sector, we find in Huerta de Valencia that agriculture maintains low employment proportion, less 2%, with progressive decrease.

- Employment per hectare of cultivation: agricultural employment in total surface of farmland.

Farms with land	Farms	AWU´s	TS (Ha)	UFS (Ha)	UTA´s/ST(Ha)	UTA´s/SAU(Ha)
Huerta de Valencia	10,030	4,542.88	21,975	20,122.77	0.21	0.23
Provincia Valencia	67,774	37,740.99	431,429	309,171.22	0.09	0.12
Comunidad Valenciana	120,180	71,020.10	890,426	657,469.61	0.08	0.11

Table 11. AWU´s/surface.

However, it must be pointed out, in **Table 11**, that agricultural employment per hectare of cultivation is most important in this area.

- Number of companies. In **Table 12** it is described the types of businesses and their legal form.

	Farmland	%	TS (Ha)	%	SAU (Ha)	%
All farms	13,321	100.00	22,761	100.00	20,410	100.00
Individual entity	12,977	97.42	19,769	86.85	18,096	88.66
Individual entity and manager	11,550	86.71	17,486	76.82	15,986	78.32
Company	97	0.73	723	3.18	662	3.24
Public entity	20	0.15	548	2.41	62	0.30
Cooperative entity	8	0.06	83	0.36	83	0.41
Agrarian Transformation Company	62	0.47	601	2.64	582	2.85
Other legal company condition	157	1.18	1,038	4.56	925	4.53

Table 12. Farmland and surface, by farmland type.

3.3.2. Cultural-scientific function

Agricultural land near cities is important for food availability. We can still have a fresh supply of food preserving agricultural land near cities.

The cultural and scientific value is recognized by agricultural activity development and local knowledge related to farmland multifunctionality and sustainable land usage:

- Scientific: related to agronomic engineering. In many cases, specific local environmental and external factors drive urban farms to develop unique innovations for space-intensive production systems, often creating a dominant paradigm for urban farming for a given location [28].
- Culture: Agricultural land near cities is an important part of cultural heritage.
- Education: Farms near cities are important places for people to learn about the natural world, and provides important opportunities to teach people about where their food comes from.

The scientific value in Huerta de Valencia is recognized by agricultural activity development throughout the centuries, from Roman times, with the development of environmental practices adopted by farmers:

- Development of crop farming that has changed over the centuries, providing plant foods, raw materials and biological material. Actually, supplying a variety of quality fresh produce,
- For each soil, farmers choose the most appropriate crop, and the best agricultural practices, creating a peculiar vegetation cover, resulting in a landscape-type non-common in Europe,
- Special working tools design adapted to the structures' type, plot size, shape of the plots, alignment of the plots, crops and agricultural practices,
- Specific vocabulary,
- Adaptation of plots for soil conservation: embankments, low walls,
- Small hydrological corrections, variety of techniques to collect and provide water to crops,
- Species conservation and preservation of plant genetic resources,

Cultural value derive from adaptation of activity, to the ecosystem:

- creating a unique space, organized according to the farming system,
- own customs, such as those derived from water resources management models through a historic hydraulic infrastructures with mills, towers, canals,
- development of agricultural practices that have contributed to soil conservation and improvement of their agrologic characteristics,
- creation of an important architectural heritage and civil engineering, with an important set of built elements (buildings, farms, barracks houses), and other elements associated with

the activity, and irrigation management, such as mills, canals, dams, etc. Glick [44], an American historian and scholar of historical irrigation systems, has studied the irrigation system in the Huerta de Valencia, and places its origins in medieval times.

Indicators:

- unique innovations
- listed as protected architectural heritage: farms, "barracas", mills, etc.
- existence of a network of historic trails
- listed as protected hydraulic heritage

    - length preserved historic canals
    - floodgate, sluice

4. Conclusion

On our objective in this paper was created a descriptive approach for the functionality of peri-urban agricultural systems, and proposed a methodology to quantify the functionality of peri-urban agricultural systems by means of indicators. Here, we reviewed the literature on per-urban agricultural system, the literature on indicators, to define an analytical framework to determine and assess ecosystem services at peri-urban agricultural systems. The review shows a wide variety of functions that can be grouped according to their economic dimension, social dimension and environmental dimension. The descriptive approach for the functionality includes a list of the goods and services that an agricultural system can provide, distinguishes between those considered private from public, and those with a positive or a negative externality nature. All goods and services are classified according to whether they are economic, social or environmental. Up to 12 generic services are defined, which are subdivided into 22 functions and 22 goods and services.

To quantify the functionality of peri-urban agricultural systems we propose 36 indicators that cover the three components of the concept (economic, social and environmental). Some of these indicators are defined from previous studies around sustainable agriculture. There are some papers which propose indicators to economic and environmental analysis but do not completely cover the selected aspects of multifunctionality. In order to yield a broader coverage, we include new indicators defined specifically for the analysis of the peri-urban agriculture particularities. Specifically, indicators related to territorial aspects of the special location of these agricultural systems, and pressures to which they are subject. Other indicators related to the education and cultural function that these agricultural systems can provide to the hipsters.

From descriptive approach for the functionality of peri-urban agricultural systems derive that these can´t be considered as an economic activity *sensu stricto*. On the other hand, society demands of these opportunities to undertake recreational activities, produce quality organic products locally, conserve and maintain spaces relatively near cities and towns under suitable environmental conditions for living, conserving cultural ethnological heritage, etc.

The planners and political decision makers should consider society demands and the contribution from peri-urban system to the economic, socio-ecological, psychological, cultural, and spiritual welfare of the urban community.

So, to planners, it is important to determine functions and services actually, in the peri-urban system. Then, it can determine social preferences for them though determining social welfare function. Finally, planners and political decision makers can define desirable future scenarios to the peri-urban agriculture system, quantifying the weight, as a starting point, and define policies to achieve those desirable scenarios.

The analytical framework presented can apply for every peri-urban system. The proposed methodology was applied to the Huerta de Valencia, a peri-urban agricultural system around Valencia City. This is a rich agricultural area with a variety of resources. There is an open political-institutional debate to define a protection scheme. The results from this study help to enrich this debate.

Indicators proposed are relevant to quantify every empirical value of function and services that explain. But the indicators proposed are highly questionable as there is no clear connection between the indicator value pure amount and the value of multifuntionality. Some authors [37] recommend transforming base indicators into adimensional variables (normalization) and then aggregate. Another question is determine the minimum/maximum values of the indicator values, a range of values that determine the multifunctional character for each aspect, or for the aggregate value.

For next research it is necessary to work in the normalization of indicators, as previous step to aggregation. To evaluate the multifunctional character it is necessary to have a range of values.

References

[1] Fielke SJ & Bardsley DK. 2015. *Regional agricultural governance in peri-urban and rural South Australia: strategies to improve multifunctionality.* Sustainability Science, Vol. 10, pp. 231–243.

[2] Renting H, Rossing WAH, Groot JCJ, Van der Ploeg JD, Laurent C, Perraud D, Stobbelaar DJ & Van Ittersum MK. 2009. *Exploring multifunctional agriculture. A review of conceptual approaches and prospects for an integrative transitional framework.* Journal of Environmental Management, pp. S112–S123.

[3] Mather A, Hill G & Nijink M. 2006. *Post-productivism and rural land use: cul de sac or challenge for theorization?* Journal of Rural Studies, pp. 441–455.

[4] Organisation for Economic Co-operation and Development (OECD). 2001. *Multifunctionality: Towards an Analytical Framework.* París: OECD Publications, p. 27.

[5] Organisation for Economic Co-operation and Development . 2003. *Multifunctionality: The Policy Implications.* París: OECD Publications.

[6] Wilson GA. 2001. *From productivism to post-productivism... and back again? Exploring the (un) changed natural and mental landscapes of European agriculture.* Transactions of the Institute of British Geographers, Vol. 26, pp. 77–102.

[7] Wilson GA. 2007. *Multifuntional Agriculture. A Transition Theory Perspective.* Wallingford: CAB International, p. 374.

[8] Jongeneel RA, Polman NBP & Slangen LHG. 2008. *Why are Dutch farmers going multifunctional?* Land Use Policy, Vol. 25, pp. 81–94.

[9] Reig E. 2002. *La multifuncionalidad del Mundo Rural.* Revista de Economía, Información Comercial Española, Vol. 803, pp. 33–44.

[10] Amekawa Y, Sseguya H, Onzere S & Carranza I. 2010. *Delineating the multifunctional role of agroecological practices: toward sustainable livelihoods for smallholder farmers in developing countries.* Journal of Sustainable Agriculture, Vol. 34, pp. 202–228.

[11] Huylenbroeck Van, Vandermeulen V, Mettepenningen E & Verspecht A. 2007. *Multifuntionality of agriculture: a review of definitions, evidence and instruments.* Living Reviews in Landscape Research.

[12] Reig E. 2007. *Fundamentos Económicos de la Multifuncionalidad [aut. libro]. Pesta y Alimentación Ministerio de Agricultura. La Multifuncionalidad de la Agricultura Española.* Madrid: Eumedia, pp. 19–39.

[13] Wilson GA. 2008. *From 'weak' to 'strong' multifunctionality: conceptualising farm-level multifunctional transitional pathways.* Journal of rural studies, Vol. 24, pp. 367–383.

[14] Brandt J & Vejre H. 2004. Multifunctional landscapes – motives, concepts and perspectives. *Multifunctional Landscapes, Volume I—Theory, Values and History*, Vol. 1, pp. 3–31.

[15] Zander P, Knierim A, Groot JCJ & Rossing WAH. *Multifunctionality of agriculture: tools and methods for impact assessment and valuation*. Agriculture, Ecosystems & Environment, Vol. 120, pp. 1–4.

[16] Compés RG & Álvarez-Coque JM. 2009. *La reforma de la PAC y la*. Opex. Observatorio de política exterior. Fundación Alternativas.

[17] Chiara M, Stefano C & Guido S. 2015. *Agricultural land consumption in periurban areas: a methodological approach for risk assessment using artificial neural networks and spatial correlation in Northern Italy*. Applied Spatial Analysis and Policy, pp. 1–18.

[18] Rogge E & Dessein J. *Perceptions of a small farming community on land use change and a changing countryside: a case-study from Flanders*. European Urban and Regional Studies, pp. 1–16. DOI: 10.1177/0969776412474664

[19] Filippini R, Marraccini E, Lardon S & Bonari E. 2014. *Assessing Food Production Capacity of Farms in Periurban Areas*. Italian Journal of Agronomy, Vol. 9.

[20] S Corsi, Mazzocchi C, Sali G, Monaco F & Wascher D. 2015. *L'analyse des systèmes alimentaires locaux des grandes métropoles. Proposition méthodologique à partir des cas de Milan et de Paris*. Cah Agric, Vol. 24, pp. 28–36.

[21] Pascucci S. 2007. *Agricoltura periurbana e strategie di svilippo rurale*. Nápoles: Dipartamento Di Economia e Politica Agraria, Università degli Studi di Napoli Federico II.

[22] Zasada I. 2011. *Multifutional peri-urban agriculture—A review of social demands and the provision of goods services by farming*. Land Use Policy, Vol. 28, pp. 639–648.

[23] Moreno O. 2009. Estratégias y Dinámicas de las Explotaciones Agrarias de Base Familiar: El caso de una Agricultura-Intensiva (thesis).

[24] Reig E. 2008. *Agricultura, desarrollo rural y sostenibilidad medioambiental*. CIRIEC, Vol. 61, pp. 103–126.

[25] Gómez-Villarino T & Gómez-Orea D. 2012. *Agricultura y Medio Ambiente: en pos del desarrollo sostenible*. Universidad de Flores, UFLO Calidad de VIda, Vol. 7, pp. 3–22.

[26] Ivesa CD & Kendal D. 2013. *Values and attitudes of the urban public towards peri-urban agricultural land*. Land Use Policy, Vol. 34, pp. 80–90.

[27] Atance I. 2007. *Política Agraria Para una Agricultura Multifuncional. Un Análisis de la PAC Reformada Frente a la Multifuncionalidad [aut. libro]. Pesca y Alimentación Ministerio de Agricultura. La Multifuncionalidad de la Agricultura Española*. Madrid: Eumedia, pp. 91–102.

[28] Pfeiffer A, Silva E & Colquhoun J. 2015. *Innovation in urban agricultural practices: responding to diverse production environments*. Renewable Agriculture and Food Systems, pp. 79–91.

[29] Pelzer E, Pelzer E, Fortino G, Bockstaller C, Angevin F, Lamine C, Moonen C, Vasileiadis V, Guérin D, Guichard L, Reau R & Messéan A. 2012. *Assessing innovative cropping systems with DEXiPM, a qualitative multi-criteria*. Ecological Indicators, pp. 171–182.

[30] Thapa RB & Murayama YM. 2008. *Land evaluation for peri-urban agriculture using analytical hierarchical process and geographic information system techniques: a case study of Hanoi*. Land Use Policy, Vol. 25, pp. 225–239.

[31] Mittenzwei K, Fjellstad W, Dramstad W, Flaten O, Loureiro M & Prestegard SS. 2007. *Opportunities and limitations in assessing the multifunctionality of agriculture within the CAPRI model*. Ecological Indicators, Vol. 7, pp. 827–838.

[32] Wiltshirea K, Delate K, Flora JL & Wiedenhoeft M. 2011. *Socio-cultural aspects of cow–calf operation persistence in a peri-urban county in Iowa*. Renewable Agriculture and Food Systems, Vol. 26, pp. 60–71.

[33] Silva R. 2010. *Multifuncionalidad agraria y territorio. Algunas reflexiones y propuestas de análisis*. EURE, Vol. 109, pp. 5–33.

[34] Laterraa P, Orúea ME & GC Boo. 2012. *Spatial complexity and ecosystem services in rural landscapes*. Agriculture, Ecosystems and Environment, pp. 56–67.

[35] Howe RW, Regal RR, Niemi GJ, DanzNP & Hanowski JM. 2007. *A probability-based indicator of ecological condition*. Ecological Indicators, pp. 793–806.

[36] Brinkley C. 2012. *Evaluating the benefits of peri-urban agriculture*. pp. 259–269.

[37] Gómez-Limón JA & Sanchez-Fernandez G. 2010. *Empirical evaluation of agricultural sustainability using composite indicators*. Ecological Economics, pp. 1062–1075.

[38] OECD. 2001. *Environmental Indicators for Agriculture Volume 3: Methods and Results*. París: OECD Publications.

[39] Brouwer F. 2002. *Effects of agricultural policies and practices on the environment: review of empirical work in OECD countries*. Directorate for Food, Agriculture and Fisheries, Organisation for Economic Co-operation and Development. COM/AGR/CA/ENV/EPOC(2001)60/FINAL.

[40] Berbel Vecino J. & Gutiérrez Martín C. 2004. *Sustainability of European Irrigated Agriculture under Water Framework Directive and Agenda 2000 – WADI*.

[41] Van Zanten BT, et al. 2014. *European agricultural landscapes, common agricultural policy and ecosystem services: a review*. Official Journal of the Institut National de la Recherche Agronomique (INRA), Vol. 34, pp. 309–3025.

[42] Muñoz F. 2003. *Lock living: urban sprawl in Mediterranean cities*. Cities, Vol. 20, pp. 381–385.

[43] de Zeeuw H. 2003. *Annotated Bibliography on Urban Agriculture*. Leusden, The Netherlands: ETC – Urban Agriculture Programme in cooperation with TUAN and other organisations, Swedish International Development Agency (Sida).

[44] Glick TF. 2003. *Regadíos y Sociedad en la Valencia Medieval*. Biblioteca Valenciana.

[45] Galli M, et al. 2010. *Agricultural Management in Peri-Urban Areas. The Experience of an International Workshop*. Ghezzano: Felici Editore Srl.

[46] Hoogevee Y, et al. 2004. *High nature value farmland. Characteristics, trends and policy challenges*. Environmental Production, European Environment Agency. Copenhagen: Scanprint a/s. ISBN 92-9167-664-0.

[47] Wilson GA. 2007. *Multifunctional Agriculture: A Transition Theory Perspective*. CABI.

Chapter 6

Relationship between Population and Agricultural Land in Amasya

Abstract

Urban development endangers agricultural and natural areas. It causes the rural population to immigrate to urban areas due to appealing life standards and leads to the extinction of rural areas. In addition, reduced rural population causes urban areas to select rural areas as development areas. This problem of rural areas can be better described in the areas that have completed their urban development but still continue to develop. This article discusses the effect of population in determining the areas which see population increases and are under pressure.

We addressed Amasya in our research. The population in the neighbourhoods of Amasya was determined. The distribution of rural areas in the city plans in the neighbourhoods was determined. Geographical information system was used for these analyses. Pearson's product-moment correlation was used on agricultural and population data obtained from the neighbourhoods. The existence of a negative or positive interaction between population and agricultural areas was shown. This study described the problems of rural areas located in urban areas and indicated their status according to the population, using statistical analysis.

Keywords: agricultural areas, urban development, geographical information system, Pearson's product moment correlation, rural population

1. Introduction

Rapid urbanization puts serious pressure on agricultural land and natural areas [1]. Urbanization has greatly reduced agricultural areas [2]. According to the OECD [3], cities are the most important sources of major problems for rural areas [4]. Uncontrolled development also reduces biodiversity and depletes natural resources. All these show that urban development endangers agricultural and natural areas.

Urban areas interact with rural areas at various levels, at regional, urban and settlement scales [5]. At the regional scale, urban areas affect rural areas in metropolitan development. At the urban scale, we can talk about the interaction between rural areas and the city at the urban fringe. Finally, at the settlement scale, the interaction between farm buildings and open spaces and the urban square is evident [5]. With this approach, urban development is threatening agricultural land at different scales. This threat plays a key role in decreasing agricultural production in these areas and increasing rapid and often uncontrolled land-use development [6, 7]. Because of decreasing agricultural areas, in the meantime, urban areas may finish all development options in agricultural areas [7].

It causes the rural population to migrate to urban areas due to appealing life standards and leads to the elimination of rural areas. In addition, reduced rural population causes urban areas to identify rural areas as development areas. The problem of rural areas can be better described in areas that have completed their urban development, and still continue to develop.

Rapid urban development brings the risk of destroying agricultural and natural areas. Controlling this development is of great importance; however, many cities were developed without taking this into consideration. Industrialization and urban development have seen many examples of this situation. The main factor is population movement. Cities with increasing population density target agricultural and natural areas as development areas.

Population movements increase urban populations and reduce rural populations. This reduces labor productivity in agricultural areas and causes these areas to remain inactive, and increases the pressure of urban development on these areas. Therefore, protecting these areas was attempted in order to reduce the pressure of urban development on the agricultural and natural areas. However, these efforts could not reduce the pressure of urban development on rural areas, since housing is a primary human need. We put land-use planning under protection, but population movements continue to threaten these planning approaches. As the general demand of population increases continues to concentrate in and around cities [2, 8], it will reduce urban agricultural areas. The roles of growing urban areas in countries where the migration of most of population is seen and land-use development of cities are expected to increase [2, 9].

It is important to analyze the relationship of urban development with population in urban agricultural areas to better understand its effect on urban agriculture, and to generate planning approaches with different points of view towards developed and developing cities. This study aims to reveal the status of urban agriculture in developing cities and the required development strategy. This article discusses the effect of population in determining the areas which see population increases and are under pressure.

2. Area description

The study area was chosen to be Amasya, located in the Central Black Sea region in the north of Turkey. Amasya is a city which borders Samsun, Tokat, Çorum and Yozgat (**Figure1**). Amasya has not completed its urban development yet, but is open to development. There are some areas of urban agriculture within or at the boundaries of the city. It is important to define these agricultural areas and determine their status within urban development.

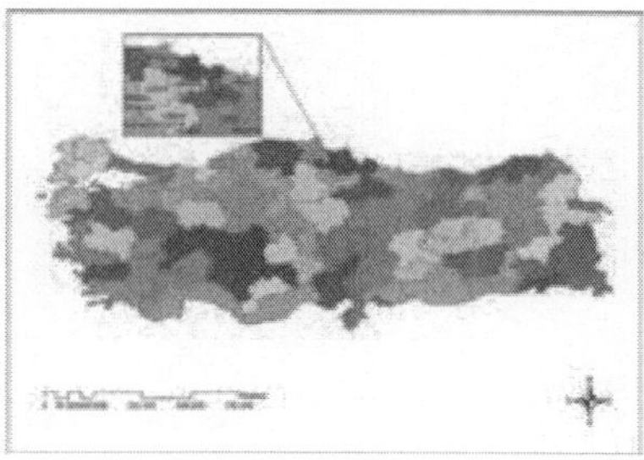

Figure 1. Study area— Amasya city (Turkey).

In addition, Amasya is a city having fertile soil and a history based on agriculture due to rich green areas and watershed characteristics. Therefore, it is an important sample for this research.

3. Material and method

According to the information on agricultural infrastructure from the Amasya Provincial Directorate of Agriculture's website, agriculture is the foundation of Amasya's economy, and 34.3% of the provincial territory consists of agricultural land [10]. In this sense, this study generates important data in terms of examining Amasya's agricultural structure and revealing the impact of urban development on agricultural land.

This study has three main steps. First, the data were organized according to Amasya's urban planning, and the agricultural land-use areas were identified. The population data of its neighbourhoods were obtained from the TurkStat [11] website. Second, the relationship between the agricultural areas and the neighbourhoods' population data was analyzed using Pearson's product-moment correlation. Third, the results of the data were evaluated and analyzed, yielding the study's results for Amasya.

Our study determined the population in the neighbourhoods of Amasya and the distribution of rural areas in the city plans. Geographical information system was used for these analyses. Pearson's product-moment correlation was used with the agricultural and population data obtained from the neighbourhoods. This study described the problems of rural areas located in urban areas and indicated their status according to the population, using statistical analysis.

The urban agricultural areas of the neighbourhoods of Amasya's development plans were analyzed (**Figure2**). The development plan was obtained from the Amasya Municipality for research. The agricultural and other areas were determined from this plan for the purpose of this study. Agricultural areas include vineyards and orchards. Cemeteries, woodlands and green areas were also shown together with other land uses; however, Pearson's product-moment correlation was only used for agricultural areas and population.

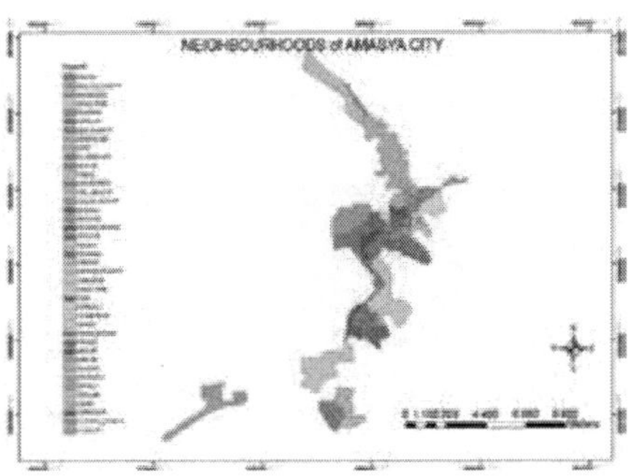

Figure 2. Neighbourhoods map of Amasya city.

The relationship between agricultural areas and population rates in the neighbourhoods was studied using correlation analysis. The product-moment correlation coefficient, which is used in calculations on an interval/ratio scale [12], was used to better understand the relationship between these data. The calculations were done to determine the effect of population growth on urban agricultural areas. Pearson's product-moment correlation formula from the book of Lee and Wong, *Statistical Analysis with Arcview GIS* (2001), was used:

$$r\sum_{i=1}^{n} xiyi / n - \overline{X}\overline{Y} / SxSy \tag{1}$$

where n shows the number of neighbourhoods, x_i and y_i show the data obtained by adding the number found by multiplying the number of agricultural areas with the population in neighbourhoods, $\overline{X} \wedge \overline{Y}$ shows the mean of agricultural areas and the population in neighbourhoods, and S_x and S_y show the standard deviations in agricultural areas and neighbourhoods. These results suggest that [12] the correlation coefficient indicates a positive relationship as it approaches +1 and a negative relationship as it approaches −1. Zero indicates a neutral relationship between the data. This analysis yielded the following results for Amasya.

4. Results and discussion

This analysis is important for understanding the pressure of the increase in urban population on rural areas. No agricultural areas remain in some of Amasya's 41 neighbourhoods. Most of the remaining agricultural areas are located on the city's boundaries. There are fewer square meters of agricultural areas in and around the city center than at the city's edges.

The agricultural areas are indicated by red on the land-use map of Amasya. There is less red color in the center of the city; however, it can still be seen at the edges (**Figure 3**). This suggests that rural areas or urban agricultural areas are gradually being reduced.

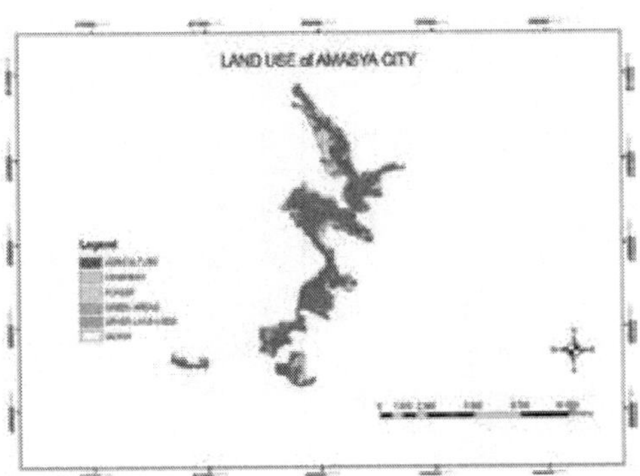

Figure 3. Land use map of Amasya city.

Table 1 indicates that the neighbourhoods with the highest ratio of agricultural land are Göllübağları and Karasenir. However, in Kirazlıdere, Savadiye, Beyazıtpaşa, Sofular, Bahçeleriçi, Pirinçci, Gümüşlü, Fethiye, Şamlar, Nergis, Dere, Hatuniye, Hacıilyas, Üçler, Kurşunlu, Gökmedrese, Şehirüstü, Yüzevler and İhsaniye, agricultural areas have been completely or almost completely destroyed. We can see the agricultural lands starting to get disappeared within the urban development of Amasya.

Layer	Agriculture	Other land uses	Forest	Cemetery	Silvan	Green areas
Karasenir	1,798,586	1,176,519	0	10,142.4	145,374.4	0
Helvaci	1,044,667	669,398.6	321,176.1	10,142.4	172,420.8	0
Cakallar	571,355.4	216,371.3	858,723.4	0	0	0
Orman_Baglari	872,246.6	135,232	0	0	0	0
Demetevler	77,758.42	50,712.01	0	0	84,520.02	0
Koza	368,507.3	270,464.1	0	0	94,662.43	0
Kapikaya_Mucavir	331,318.5	294,129.7	483,454.5	0	338,080.1	3,380.801
Gollubaglari	1,947,341	997,336.3	1,254,277	162,278.4	10,142.4	0
Sarilar	104,804.8	67,616.02	0	0	0	0
Akbilek	845,200.2	689,683.4	114,947.2	0	20,284.81	3,380.801
Ellibesevler	169,040	432,742.5	0	0	0	10,142.4

Layer	Agriculture	Other land uses	Forest	Cemetery	Silvan	Green areas
Seyhcui	716,729.8	879,008.2	341,460.9	3,380.801	74,377.62	13,523.2
Kirazlidere	3,380.801	4,83,454.5	216,371.3	0	0	0
Savadiye	0	67,616.02	0	0	30,427.21	0
Beyazitpasa	0	141,993.6	0	0	0	0
Mehmetpasa	0	87,900.82	0	0	0	3,380.801
Sofular	0	128,470.4	0	0	40,569.61	16,904
Bahcelerici	0	365,126.5	0	0	3,380.801	20,284.81
Pirincci	0	64,235.22	0	0	0	0
Gumuslu	0	13,523.2	0	0	0	0
Fethiye	0	111,566.4	13,523.2	0	0	0
Samlar	10,142.4	155,516.8	94,662.43	6,761.602	13,523.2	0
Nergis	0	67,616.02	3,380.801	0	0	3,380.801
Dere	0	138,612.8	0	0	0	0
Hatuniye	0	138,612.8	10,142.4	0	0	0
Haciilyas	0	118,328	0	0	0	0
Ucler	0	64,235.22	0	0	0	0
Kursunlu	0	179,182.4	47,331.21	0	0	0
Hacilar_Meydani	250,179.3	571,355.4	0	0	6,761.602	3,380.801
Yazibaglari	828,296.2	273,844.9	0	3,380.801	0	0
Gokmedrese	0	121,708.8	463,169.7	43,950.41	30,427.21	0
Sehirustu	0	54,092.81	3,380.801	0	0	0
Yuzevler	0	74,377.62	0	0	0	0
Hizirpasa	50,712.01	429,361.7	57,473.62	0	0	50,712.01
Findikli	348,222.5	317,795.3	0	3,380.801	135,232	0
Bogazkoy	128,470.4	101,424	57,473.62	0	0	0
Ihsaniye	0	182,563.2	108,185.6	0	6761.602	0
Gozlek_Mucavir	402,315.3	30,427.21	0	0	0	0
Yildizkoy_MucavI	57,473.62	40,569.61	81,139.22	0	0	0
Baglica_Mucavir	101,424	37,188.81	104,804.8	0	0	0
Dadi_Mucavir	561.213	294.129.7	0	0	0	0

Sources: The data on land-use areas were obtained and designed from the land-use map that was given by the Municipality of Amasya.

Table 1. Data of land uses in Amasya city.

Table 1 shows that urban development is suppressing agricultural areas in Şeyhcui, Hacılar Meydanı and Hızırpaşa neighbourhoods. Moreover, in Akbilek, Karasenir Dadı_Mucavir Area and Fındıklı and Kapıkaya_Mucavir Area, other land uses are dominating the neighbourhoods. Here, other land uses indicate urban infrastructure. It is evident that the city of Amasya has a development-oriented approach, which results in urban development pressure on existing agricultural areas.

In order to make better sense of the available data, Figure 3 should be examined in table format. Accordingly, Table 1 shows the 41 neighbourhoods of Amasya city and the land-use.

As Table 1 shows, other land uses, which indicate urban land uses, exist variably in every neighbourhood. This can be considered to be an indicator that even rural areas of the Amasya city are becoming involved in the urban development process. Therefore, Amasya needs an urban development approach that could protect its agricultural land.

The effect of population growth on rural areas in developed or developing urban areas can be seen, and urban agricultural areas are disappearing in the city. The status of rural areas confined by the city will be indicated, depending on the population movements. The effect of urban development on these areas is even stronger here. Measures for the protection of rural areas from urban development will be determined according to the data provided (**Table 2**) in the conclusion.

$$n = 41$$

$$\sum x = 98822 \sum y = 11589385 \cdot 51$$

$$\overline{X} = 2410 \cdot 292683 \ \overline{Y} = 282667 \cdot 9393$$

$$\sum x^2 = 755938164 \sum y^2 = 12104402445096 \cdot 600$$

$$\overline{X^2} = 5809510 \cdot 817 \ \overline{Y^2} = 79901163902$$

$$Sx = 3553.590491; Sy = 464034.6582$$

$$r = -0.40$$

Neighbourhoods in Amasya	Total popu lation	Total agricultural area	X^2	Y^2	XY
Karasenir	789	1,798,586.083	622,521	3,234,911,897,961.280	1,419,084,419
Helvaci	1.996	1,044,667.48	3,984,016	1,091,330,143,769.550	208,515,6290
Cakallar	403	571,355.3534	162,409	326,446,939,858.838	230,256,207.4
Orman_Baglari	165	872,246.6341	27,225	760,814,190,698.7790	14,392,0694.6

Neighbourhoods in Amasya	Total popu lation	Total agricultural area	X^2	Y^2	XY
Demetevler	304	77,758.42087	92416	6,046,372,016.19605	23,638,559.94
Koza	1.330	368,507.2989	1,768,900	135,797,629,342.5740	490,114,707.5
Kapikaya_Mucavir	484	331,318.4889	234,256	109,771,941,086.9790	160,358,148.6
Gollubaglari	817	1,947,341.323	667,489	3,792,138,228,263.3900	159,097,7861
Sarilar	443	104804.8281	196,249	10,984,051,993.0705	46,428,538.85
Akbilek	6.624	845,200.2269	43,877,376	714,363,423,551.8120	5,598,606,303
Ellibesevler	8.946	169,040.0454	80,030,916	28,574,536,948.8341	1,512,232,246
Seyhcui	16.252	716,729.7924	2,641,27504	513,701,595,313.7470	11,648,292,586
Kirazlidere	2.934	3,380.800907	8,608,356	11,429,814.7728	9,919,269.861
Savadiye	1.152	0	1,327,104	0.0000	0
Beyazitpasa	2.984	0	8,904,256	0.0000	0
Mehmetpasa	1.847	0	3,411,409	0.0000	0
Sofular	519	0	269,361	0.0000	0
Bahcelerici	11.253	0	12,663,0009	0.0000	0
Pirincci	497	0	247,009	0.0000	0
Gumuslu	193	0	37,249	0.0000	0
Fethiye	1.476	0	2,178,576	0.0000	0
Samlar	1.394	10142.40272	1,943,236	102,868,332.9347	14,138,509.39
Nergis	238	0	56,644	0.0000	0
Dere	806	0	649,636	0.0000	0
Hatuniye	168	0	28,224	0.0000	0
Haciilyas	1.941	0	3,767,481	0.0000	0
Ucler	1.691	0	2,859,481	0.0000	0
Kursunlu	2.121	0	4,498,641	0.0000	0
Hacilar_Meydani	12.003	250,179.2671	1,440,72009	62,589,665,686.6931	300,290,1743
Yazibaglari	2.021	828,296.2223	4,084,441	686,074,631,876.4510	167,398,6665
Gokmedrese	1.100	0	1,210,000	0.0000	0
Sehirustu	2.082	0	4,334,724	0.0000	0
Yuzevler	1.492	0	2,226,064	0.0000	0
Hizirpasa	5.774	50,712.01361	3,333,9076	2,571,708,324.3808	292,811,166.6
Findikli	744	348,222.4935	553,536	121,258,904,979.3580	259,077,535.2
Bogazkoy	795	128,470.4345	632,025	16,504,652,540.6188	102,133,995.4
Ihsaniye	1.988	0	3,952,144	0.0000	0

Neighbourhoods in Amasya	Total popu lation	Total agricultural area	X^2	Y^2	XY
Gozlek_Mucavir	199	402,315.308	39,601	161,857,607,051.1350	80,060,746.29
Yildizkoy_Mucavir	163	57,473.61543	26,569	3,303,216,470.5955	9,368,199.315
Baglica_Mucavir	445	101,424.0272	198,025	10,286,833,293.4663	45,133,692.1
Dadi_Mucavir	249	561,212.9506	62,001	314,959,975,921.1580	139,742,024.7

Sources: The data of agricultural areas were obtained and designed from the land-use map given by the Municipality of Amasya. The data of population was obtained from the Turkstat [11] website (https://biruni.tuik.gov.tr/adnksdagitapp/adnks.zul).

Table 2. Data for Pearson's product-moment correlation coefficient.

A negative relationship was found between the agricultural areas and population data. The minus value indicates an inverse proportion, that is, the population falls if agricultural areas increase, and vice versa. This relationship shows that the increase in urban population leads to the destruction of urban agricultural areas. If urban agricultural areas and the increase in population cannot be balanced, the urban agricultural areas will be destroyed or eliminated by urban development.

Amasya's development continues, and its urban agricultural areas still exist. However, agricultural areas have decreased due to the increase in urban population. There are no agricultural areas in some neighbourhoods, indicating that urban development has rapidly increased and that the population in rural areas has started to replace the population in urban areas (**Figures 4** and **5**).

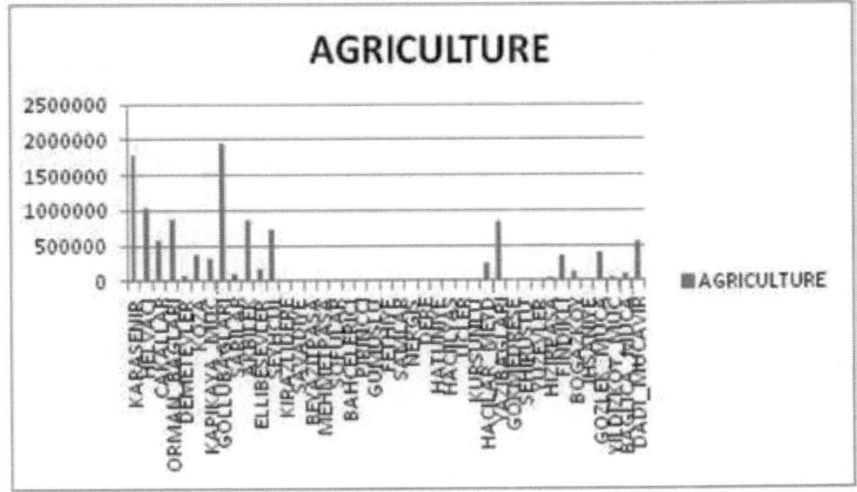

Figure 4. Land use of agricultural areas in Amasya.

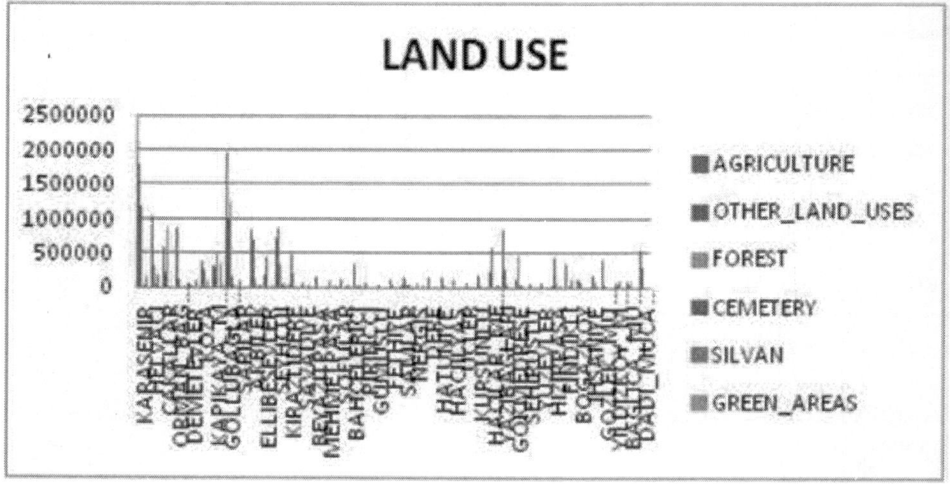

Figure 5. Land use areas in Amasya.

In conclusion, urban development targets rural and natural areas as development areas. It is inevitable that agricultural areas within urban areas be eliminated in time if they lose their special importance in the functions of the city. As a result of this study, Amasya can be an example of the cities that have not completed their urban development due to the inclusion of rural areas. The most important factor is the attempt to develop an approach to protect urban agricultural areas or to determine how to adapt these areas to the economic movements within the scope of urban development in Amasya. If agriculture comes to the fore as an indispensable factor for Amasya, where industrial development has not increased much, then Amasya will be a city that would not lose, but rather protect its rural areas.

Rural areas are usually concentrated in locations with ample water, transportation and production. The best precaution against wasting agricultural land in the urban development process is defining agricultural lands and directing urban development towards areas without agricultural value. Moreover, agricultural incentives and support for farming families should be prioritized.

References

[1] Pribadi DO, Pauleit S. The dynamics of peri-urban agriculture during rapid urbanization of Jabodetabek metropolitan area. Land Use Policy. 2015;48:13–24. DOI: http//:dx.doi.org/10.1016/j.landusepol.2015.05.009

[2] Tsuchiya K., Hara Y., Thaitakoo D. Linking food and land systems for sustainable peri-urban agriculture in Bangkok metropolitan region. Landscape and Urban Planning. 2015;143:192–204. DOI: http//:dx.doi.org/10.1016/j.landurbanplan.2015.07.008

[3] OECD, editor. Cities and Climate Change. 1st ed. OECD; 2010. 276 p. DOI: http//:dx.doi.org/10.1787/9789264091375-en

[4] Indraprahasta GS Potential of urban agriculture development in Jakarta. The Third International Conference on Sustainable Future for Human Security SUSTAIN (2012). Procedia Environmental Sciences. 2013;17:11–19. DOI: 10.1016/j.proenv.2013.02.006

[5] Torreggiani D., Dall'Ara.E, Tassinari P. The urban nature of agriculture: biodirectional trends between city and countryside. Cities. 2012;(29):412–416. DOI: 10.1016/j.cities. 2011.12.006

[6] Overbeek MMM, Terluin IJ, editors. Rural Areas Under Urban Pressure; Case Studies of Rural–Urban Relationships. LEI Wageningen UR, The Hague: Agricultural Economics Research Institute (LEI); 2006. 249 p.

[7] Aubry C., Ramamonjisoa J., Dabat M-H, Rakotoarisoa J., Rakotondraibe J., Raberharisoa L. Urban agriculture and land use in cities: an approach with the multi-functionality and sustainability concepts in the case of Antananarivo (Madagascar). Land Use Policy. 2012;(29):439–439. DOI: 10.1016/j.landusepol.2011.08.009

[8] United Nations. World Urbanization Prospects: The 2011 Revision [Internet]. 2012. Available from: http://esa.un.org/unpd/wup/index.htm [Accessed: 20.03.2014]

[9] Seto KC, Fragkias M., Güneralp B., Reilly MK Meta-analysis of global urban land expansion. PloS One. 2011;(6):e2377. DOI: http://dx.doi.org/10.1371/journal.pone. 0023777

[10] Amasya Provincial Directorate of Provincial Food Agriculture and Livestocks. Amasya'da Tarım [Internet]. Available from: http://amasya.tarim.gov.tr/Menu/10/Amasya-da-Tarim [Accessed: 31.01.2016]

[11] Turkstat. Adrese Dayalı Nüfus Kayıt Sistemi [Internet] 2015. Available from: https://biruni.tuik.gov.tr/adnksdagitapp/adnks.zul [Accessed: 15.11.2015]

[12] Lee J., Wong DWS, editors. Statistical Analysis with ArcView. United States of America: John Wiley & Sons; 2001. 192 p.

Chapter 7

Water Quality Modeling and Control in Recirculating Aquaculture Systems

Abstract

Nowadays, modern aquaculture technologies are made in recirculating systems, which require the use of high-performance methods for the recirculated water treatment. The present chapter presents the results obtained by the authors in the field of modeling and control of wastewater treatment processes from intensive aquaculture systems. All the results were obtained on a pilot plant built for the fish intensive growth in recirculating regime located in "Dunarea de Jos" University from Galati. The pilot plant was designed to study the development of various fish species, starting with the less demanding species (e.g. carp, waller), or "difficult" species such as trout and sturgeons (beluga, sevruga, etc.).

Keywords: Recirculating aquaculture system, Modeling and control, Water quality, Trickling biofilter, Expert system

1. Introduction

The recirculating aquaculture systems (RASs) became an essential component of the modern aquaculture [1–3]. The accelerated developing of RASs, which tend to become predominant with respect to the "flow-through" systems from the classic fishpond aquaculture, was stimulated by the necessity to locate the production units close to the markets, i.e. in the areas with high population density.

Thus, RASs became an important component of the *Urban Agriculture*. But the close proximity of the production centers by the sale units is just one of the advantages of RASs. Among other advantages of RASs, some even more important than the mentioned one, are the following:

- the possibility to control physicochemical parameters of the culture medium: dissolved oxygen concentration of the water, concentrations of the harmful substances (ammonia, nitrites, nitrates, carbon dioxide etc.), pH, temperature etc.;
- saving water resources. In the classical "flow-through" systems, the specific water consumption is about 10 (m^3 water/kg of fish), whereas in RASs only 5–10% of the total volume of the recirculated water is replaced with fresh water, resulting a consumption of about 0.1 (m^3 water/kg of fish);
- the possibility to control the hygienic and sanitary state of the culture biomass by removing the possibility of pathogens penetration inside RAS, applying preventive measures for diseases, the prompt achievement of the treatments when the diseases occur etc.
- providing a performant technological management concerning the populating of aquaculture tanks (i.e. populating density) for different ages of the fish biomass, implementing the feeding technology; and
- reduce the negative impact on the environment through specific means of collecting the residual solids and respecting the requirements concerning the water exhausted from RASs and discharged in the collecting urban network.

Besides the advantages mentioned above, RASs also have some drawbacks, the most important being the required investments for the equipment. Some of these—such as those for monitoring and control—are expensive. Relatively high electricity consumption to provide the water recirculating in an aquaculture system could also be mentioned.

The biological filtering process of the recirculated water has a crucial importance in RAS technology. The degree of RAS intensity, which means the ratio (fish production/space unit of culture) to provide a correct hygienic and sanitary state of the fish biomass, depends on the performance of this process. Therefore, the issue of modeling the biological filtering process is treated in this chapter with priority.

In the fish intensive growth tanks, an aerobic process takes place. The organic substances existing in the water (dejections, unconsumed food) are decomposed by heterotrophic bacteria in simpler organic products, resulting ammonia as a final product. The ammonia is also a metabolism product of fish, being released mainly by gills. However, the amount of ammonia from an aquaculture tank mostly depends on the food rate of the fish biomass. In the aquaculture tanks, the ammonia is found in two forms: the ionized form and the unionized one. The unionized ammonia is extremely toxic for the fish, and its concentration depends on the water pH and temperature.

The ammonia removal takes place through a biological filtering process that develops in two phases: (1) ammonia is oxidized by *Nitrosomonas* bacteria and transformed in nitrites, which are highly toxic and (2) the nitrites are oxidized by another category of autotrophic bacteria (*Nitrobacter*) and transformed into nitrates. The two oxidizing processes should be followed by a denitrification process, which leads to the conversion of nitrates into gaseous nitrogen. Denitrification can be achieved by either chemical or biological means. The second possibility consists in using of aquatic plants for which the nitrate is a food source enabling to achieve an aquaponic system. This is a recirculating system that provides simultaneously the fish and plant growth (usually vegetables) using a single input: fish fodders. The fish component of the aquaponic recirculating system provides the food (nitrate) for the horticultural biomass and the plants contribute through denitrification to the recirculated water purity in aquaculture tanks.

The next sections briefly present some results regarding the modeling and control of a pilot plant from "Dunarea de Jos" University of Galati consisting in a RAS with a chemical denitrificator. The next section describes the pilot plant including the technological and control equipment. Section 3 presents the mathematical model of RAS, focusing on the biological filtering processes. Some experimental results concerning the control of RAS and the possibilities of using expert systems in this purpose are included in Sections 4 and 5, respectively. The work ends with a brief section of conclusions.

2. The experimental plant

The experimental plant is located in the Intensive Aquaculture Laboratory at "Dunarea de Jos" University of Galati, Romania. It consists of two subsystems: the technological equipment and the one for monitoring and control purpose.

2.1. The technological equipment

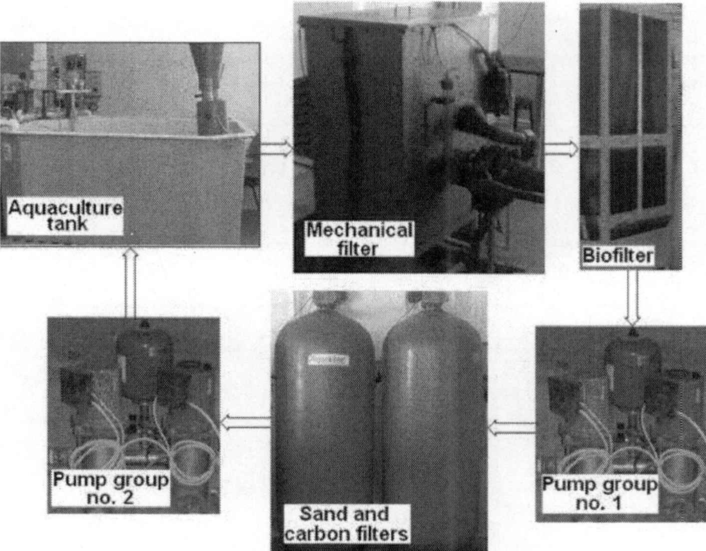

Figure 1. Structure of the technological plant.

Figure 1 shows the technological plant. It contains the following components: four aquaculture tanks of 1 m^3 each, a drum filter for rough solids removal, a collecting tank, a sand filter and an activated carbon filter for the removal of fine solids in suspension, a biological filter of trickling type together with a second collecting tank, a denitrificator that retains the nitrates, an UV filter, that acts as a disinfectant for killing the pathogenic bacteria, and the feed dosing mechanism. The aquaculture plant is also provided with an air supplying system aiming to ensure the necessary dissolved oxygen concentration in the fish tanks and in the biofilter.

2.2. Monitoring and control equipment

Figure 2 shows the monitoring and control system of RAS. It contains two control levels: the first level includes the basic control loops together with the data acquisition system; the second level has two components: an expert system for diagnosis and global control of RAS and the Human–Machine Interface (HMI).

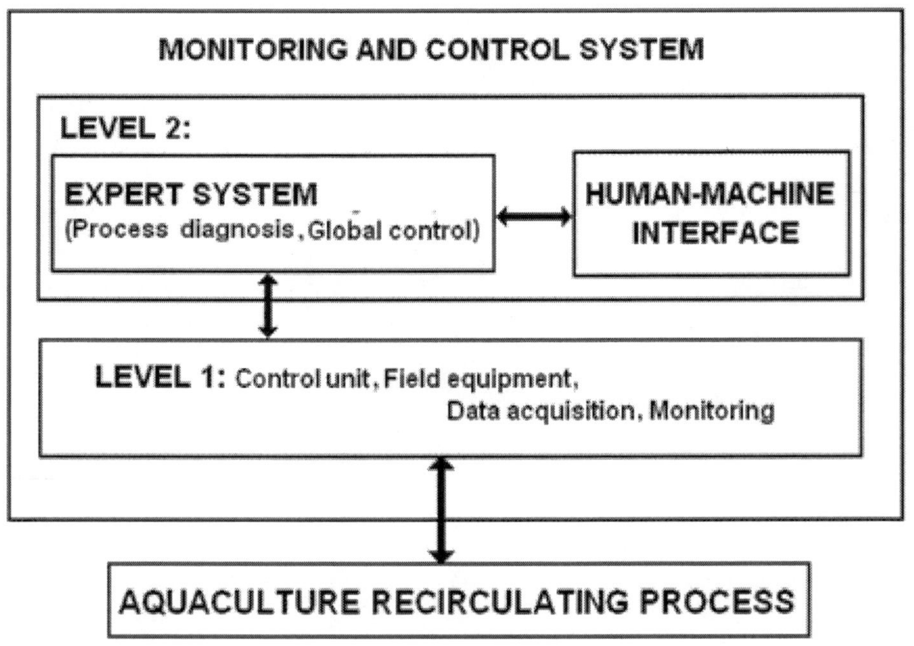

Figure 2. Monitoring and control system of recirculating aquaculture system.

Figure 3 shows the recirculating aquaculture process and the field equipment [4]. Two main circuits can be observed: a water circuit (blue) and an air circuit (red). The following field equipment can be noticed:

- Transducers: temperature (T1, T4, T7, T10 and T17); dissolved oxygen concentration (T2, T5, T8 and T11); water level in aquaculture tanks (T3, T6, T9 and T12); water level in the collecting tank located under the biofilter (T18); water flow (T13, T23–T26); pH (T15 and T20); ammonia concentration (T14 and T19); nitrate concentration (T21); nitrite concentration (T22).
- Actuators: electro-valves for air supplying control of the four aquaculture tanks (R1–R4); electro-valve for air supplying control of the trickling biofilter (R5); electro-valves for water supplying control of the four aquaculture tanks (R6–R9); pumps used for the pH control in the first collecting tank placed after the drum filter (one is for acid supply and the second is for base supply).

Another two pumps provide the necessary flow of the recirculated water within the intensive aquaculture plant. The first pump transfers the water from the drum filter to the sand and activated carbon filters and the second supplies the four aquaculture tanks with clean water taken from the biological filter.

The signal acquisition and the basic control loops are performed by a programmable logic controller (PLC), which is configured in accordance with the monitoring and control application of RAS. It communicates wireless with a computer in which the two software components, HMI and the expert system for process diagnose and global control of RAS, are implemented.

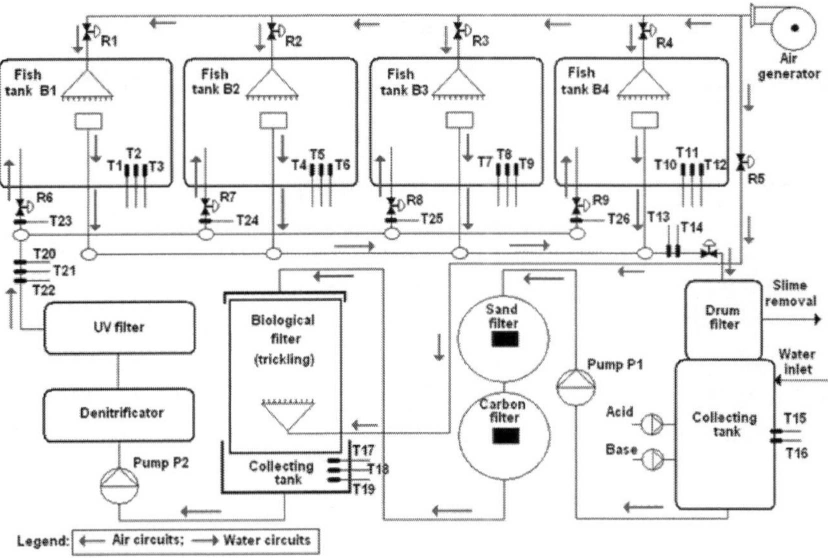

Figure 3. Experimental plant of the recirculating aquaculture system [4].

3. Mathematical modeling of intensive recirculating aquaculture systems

RAS contains three subsystems, which must be modeled: the biological system that means of culture biomass developing, the microbiological system that means of water quality and the recirculating hydraulic system that means the physical plant for water recirculating. The three subsystems have different time constants from a few minutes in the case of hydraulic system to several weeks in the case of biological system. The processes of interest, which will be approached further, are the biological process and, especially, the microbiological one. This is because the two subsystems mentioned above strongly influence the water quality, which is an essential factor for urban agriculture.

3.1. Mathematical modeling of the tanks for the growth of the fish biomass

Mathematical modeling of the tanks for the fish biomass growth involves two essential aspects:

- the model should provide information concerning the fish biomass which is in the aquaculture tanks at a given moment and the growth rate of the fish biomass. This is important to allow the calculus of the daily food ratio necessary for the proper development of the fish biomass and the estimation of the food percent assimilated by the fish biomass;
- the model should also provide information about the manner of residuals producing in aquaculture tanks. Thus, the production and consumption processes of the biochemical components of food (proteins, fat, carbohydrates, ash and water) should be considered among the types of processes occurring in the fish material: feeding, food digestion, mass growth and maintenance.

In order to estimate the fish biomass, the literature recommends two main models: using specific growth rate (SGR) or thermal growth coefficient (TGC). The second model is more advantageous compared with the use of SGR, because a very important factor of the fish biomass growth is taken into consideration: the temperature. In these conditions, the model which uses TGC will be considered for the fish biomass growth. At the same time, the model of the fish biomass growth should offer an estimation of the fish number in aquaculture tanks. These models are available between two weighing, therefore for a period of about 30 days. Based on the information about the growth rate of individual mass and the number of individuals from aquaculture tanks, the necessary daily food is determined through the feed conversion ratio (FCR).

In the modeling of the residual producing processes in aquaculture tanks, the purpose for which it is desired to build the model should be considered: achieving a global model of aquaculture plant. Thus, the model should be compatible from the state variables point of view with the model of the trickling biofilter. Therefore, it is necessary to determine a model having the following state variables: ammonia, inert components and dissolved oxygen. It starts from food decomposition in the main components: nitrogen, carbon and phosphorus. The food is introduced into aquaculture tanks in batch mode (1–2 times/day) or continuously. In the present study, taking into account that most of the plants are provided with discontinuous feeding, including the pilot plant from "Dunarea de Jos" University of Galati, it used the assumption that the food is given in batch mode. The second step is to describe how these components are affected in the main processes that are related to the food of fish biomass: feeding, digesting food, mass growth and maintenance.

The two levels of the model interact as follows: information about the growth of the fish biomass determines the food amount introduced into aquaculture tanks. This is the input information of the residual producing.

The model TGC takes also into consideration the water temperature in the body mass growth of fish biomass [5]:

$$\mathrm{TGC} = \left[\left(\mathrm{MI}_f^{1/3} - \mathrm{MI}_0^{1/3}\right) / \left(T \times t\right)\right] \times 1000 \tag{1}$$

where T is the water temperature (°C), and t is the evolution time (days).

Mass changing during a period of the temperature evolution on days ($T \times t$) is given by the following equation:

$$\mathrm{MI}(t) = \left[\mathrm{MI}_0^{1/3} + \mathrm{TGC} \times \left(T \times t\right) / 1000\right]^3 \tag{2}$$

The derivative of Equation (2) leads to obtaining the individual body mass of the fish material:

$$\mathrm{CMI}(t) = 3 \times \mathrm{TGC} \times T \cdot \left[\mathrm{MI}_0^{1/3} + \mathrm{TGC} \times \left(T \times t\right) / 1000\right]^2 / 1000 \tag{3}$$

To determine the mass of the fish material from aquaculture tanks, it is also necessary to model the evolution of the fish number during a production cycle. Thus, it is considered that number of individuals decreases with the age increase, the decrease being modeled through the decay coefficient [5]:

$$k = -1 / t_{\mathrm{CP}} \cdot \ln\left(1 - p_{\mathrm{M}} / 100\right) \tag{4}$$

where k is the decay coefficient, t_{CP} is the duration of the production cycle expressed in days, and p_{M} is the decay percent considered for the respective production cycle.

The number of individuals evolves along a production cycle accordingly to the equation:

$$n(t) = n(0) e^{-k \cdot t} \tag{5}$$

where $n(0)$ is the initial number of fishes.

In these conditions, the total fish mass can be estimated at each moment of time. The mass growth of the fish material can be determined through the derivative of the equation of total fish mass, resulting [5]:

$$\mathrm{CM}(t) = n(t)\big(\mathrm{CMI}(t) - k \times \mathrm{MI}(t)\big) \tag{6}$$

Figure 4 shows the evolutions of individual body mass (a) and the number of individuals (b) when a 140-day production cycle is considered, compared with the experimental data collected from RAS.

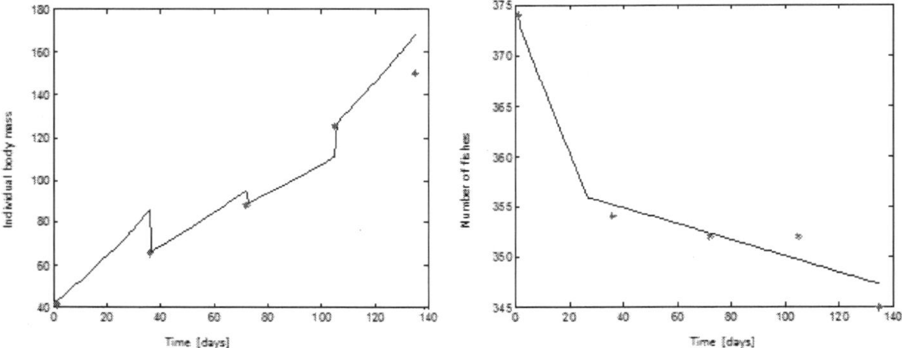

Figure 4. (a) Evolution of the individual body mass and (b) evolution of the number of individuals. Note: * = experimental data; solid line = model results.

For modeling the process of residuals producing by the fish biomass, the following four processes should be considered:

- feeding process: the food is introduced into aquaculture tanks in batch or continuous mode. The most part of food is consumed by fish, while a small fraction is lost in water;
- food digestion: after fish feeding, the amount of residuals from water increases reaching a maximum and then decreases monotonically. This process can be modeled as two first-order systems with delay, connected in series. Practically, it shows how the food is digested by the fish biomass and transformed into residuals;
- growth: this process assumes the existence of a consumption of the main elements introduced by food. The consumption is calculated in relation with the mass growth of the fish material;
- maintenance: the process determines a consumption of some elements, proportional to the total mass of fish.

The modeling of the residual producing process by the fish biomass starts from the biochemical composition of food. A typical composition of food is given in **Table 1**. Thus, for the calculus of nitrogen amount introduced through the food, it results: $N_{food} = 0.44 \times 0.16 = 0.064$ kg N/kg of food. It is considered that the food is given 2 times/day (at 6 AM and 6 PM) and the food introduced into aquaculture tanks is expressed by a function $f(t)$.

Element	Food (%)	COD (kg COD)	N (kg N)	P (kg P)
Protein	44	1.45	0.16	–
Carbohydrates	14	1.10	–	–
Fat	24	2.14	–	–
Ash	8	–	–	0.2
Water	10	–	–	–

Table 1. Biochemical composition of the food.

The food digested by the fish biomass is calculated as follows: $\tilde{f}(t)=L^{-1}\{G(s)\}\times f(t)$, where $L^{-1}\{\cdot\}$ is the inverse Laplace transformation, and $G(s)$ is the transfer function of the model of the food digestion [5]. This function will be used to determine the component of the unconsumed food lost in water $f(t) \cdot \varepsilon_p$ and the rate of residual discharge after digestion $\tilde{f}(t)\cdot(1-\varepsilon_p)$, where ε_p is the ratio of the unconsumed food. In order to determine the consumption of the main elements introduced through the food for the mass growth of the fish, the signal $\delta_T(t)$ is considered (see Figure 5a). It means the graph of the modified feeding flow to obtain a function whose area in 1 day is equal to 1. Based on the signal $\delta_T(t)$ and the digestion model, the rate of discharge corresponding to the signal $\delta_T(t)$ is obtained: $s_F(t) = L^{-1}\{G(s)\} \times \delta_T(t)$. It is plotted in Figure 5b.

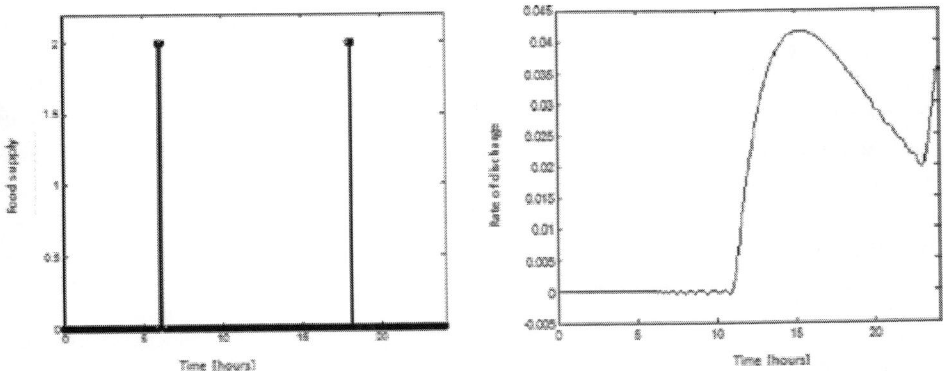

Figure 5. (a) Food supply of aquaculture tanks and (b) the evolution of the rate of discharge after digestion for 1 day.

Table 2 presents the matrix of residual producing, where the nitrogen (N) and inert substrate/biomass components (I) are highlighted. The maintenance process was not presented in Table 2 because it contributes only to the dissolved oxygen consumption without to affect other components considered in the model. The residuals production from aquaculture tanks is based on the **Table 2** and is given for each component by the sum of the following products: + Column 1 × $f(t)\cdot\varepsilon_p$ + Column 2 × $\tilde{f}(t)\cdot(1-\varepsilon_p)$ – Column 3 × $s_F(t)$ × CM(t) – Column 4 × $s_F(t)\cdot M(t)$ [5].

Residuals producing Variable	Feeding (kg of res./kg of food)	Digested food (kg of res./kg of food)	Mass growth (kg res./kg of fish/day)
S_{ND} —biodegradable soluble organic nitrogen	$0.5N_{hrana}$	$0.15N_{hrana}$	$-0.15N_{peste}$
X_{ND} —particles of biodegradable organic nitrogen	$0.5N_{hrana}$	$0.15N_{hrana}$	$-0.15N_{peste}$
S_{NH4} —ammonia	0	$0.7N_{hrana}$	$-0.7N_{peste}$
X_I —inert biomass	$0.5I_{hrana}$	$0.5I_{hrana}$	$-0.5I_{peste}$
S_I —inert substrate	$0.5I_{hrana}$	$0.5I_{hrana}$	$-0.5I_{peste}$

Table 2. Matrix of residuals producing.

3.2. Mathematical modeling and analysis of trickling biofilter

A biofilter of trickling type is composed by numerous vertical distributed solids which offer a large contact surface with the water that should be treated through the nitrification process. The biofilms are formed on each element of the filter, at a microscopic scale, carrying out the nitrification process. Two spatial coordinates intervene in the biofilter model: a spatial coordinate related to the biofilter height, z, corresponding to the processed water path, and a second spatial coordinate related to the biofilter thickness, ζ, corresponding to the processes from the biofilm. Taking into account the fact that the inert medium whereon the microorganisms are fixed, forming the biofilm, is not flooded, but it has wet surface and is aerated, it results that three zones which need to be modeled can be considered: the biofilm zone, the liquid zone (wastewater pellicle) and the gaseous zone. Furthermore, the flow of substance from gas to biofilm is considered null and only the biofilm and liquid zones will be modeled from the transfer of the components contained in the wastewater point of view. The gaseous zone will contribute only to the aerating process of the biofilm.

In what follows, the fundamental equations of the concentration of one component (ammonia, nitrate etc.) are considered in the biofilm and the liquid volume.

The model of concentration in the biofilm is [6]:

$$\frac{\partial c}{\partial \tilde{t}} = \frac{\partial^2 c}{\partial \xi^2} - r(c) \tag{7}$$

where c is the concentration of the component considered, ξ is the spatial coordinate related to the biofilter thickness, and $r(c)$ is the consumption rate of the component c. The spatial coordinate ξ is scaled: $\xi = \zeta/L$, where L is the biofilter thickness and $0 < \xi < 1$. The time is also scaled, $\tilde{t} = \lambda t,\ \ \lambda = D/(L^{\ 2}\varepsilon)$, where D is the diffusion coefficient, and ε is the biofilm porosity (m^3/m^3).

The boundary conditions of Equation (7) are:

$$\left(\frac{\partial c}{\partial \xi}\right)_{\xi=0} = 0\ ;\ \ (c)_{\xi=1} = c^b \tag{8}$$

where c^b is the concentration in the liquid volume.

The model of the concentration in the liquid volume is [6]:

$$v\frac{\partial c_i^b}{\partial t} = q\frac{\partial c_i^b}{\partial z} + aJ_{f,i} + a_g J_{g,i},\quad i = 1,2,\ldots,n \tag{9}$$

in which

$$v = V/(A_r h);\quad q = Q/A_r;\quad a = A/(A_r h);\quad a_g = A_g/(a_r h) \tag{10}$$

where c_i^b is the concentration of component i in liquid, $J_{f,i}$ is the flow of substance from the gas to biofilm, z is the spatial coordinate along the length of biofilter, A is the total area of biofilter, V is the total volume of liquid, A_g is the total area of the gas–liquid interface, A_r is the section area of the biofilter, h is the biofilter height, Q is the liquid flow which crosses the biofilter.

The flow from the biofilm to liquid, $J_{f,i}$, is expressed through the equation [6]:

$$J_{f,i} = -D_i\left[\frac{\partial c_i}{\partial \xi}\right]_{\xi=1} \tag{11}$$

where D_i is the diffusion coefficient for the component i.

In Equation (10), the spatial coordinate z is discretized in N finite zones which corresponds to the approximation of the distributed system model with respect to z by N concentrated parameter subsystems, connected in series, as shown in **Figure 6** (gaseous zone is considered to be common) [7]. At the level of each concentrated subsystem from **Figure 6**, the mass balance equation of the component considered has the following general form [6]:

$$V\frac{dc^b}{dt} = Q\left(c_{in}^b - c^b\right) + AJ_f + A_g J_g \tag{12}$$

where V is the liquid volume in the finite element of the subsystem, c^b is the component concentration in this finite element and c_{in}^b is the component concentration at the input of the finite element.

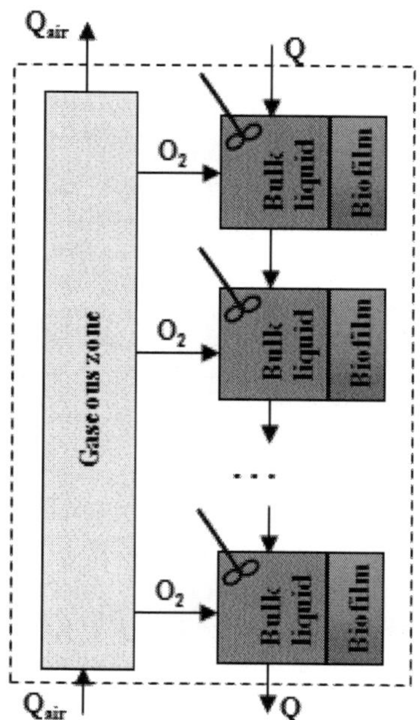

Figure 6. Structure of trickling biofilter [7].

Considering that the material flow from gas to biofilm is null and taking into account (11), Equation (12) can be written in the non-dimensional form [6]:

$$\tau\frac{dc^b}{d\tilde{t}} = c_{in}^b - c^b - \gamma\left[\frac{\partial c}{\partial \xi}\right]_{\xi=1}, \text{ with } \tau = \frac{V}{Q}\lambda, \quad \gamma = \frac{AD}{QL} \tag{13}$$

It is considered that the general model of biofilter is given by N equations of (13) form, for which every finite element resulted from the discretization of spatial coordinate, z, and N

equations of (7) form, these must offer the factor $\left[\frac{\partial c}{\partial \xi}\right]_{\xi=1}$ that intervenes in Equation (13) of each zone defined along the biofilter height.

Furthermore, the biofilter simulation through the model discretization was carried out, first of all considering the linear model of the concentration in biofilm.

If the substrate concentration is low, Equation (7) can be approximated by the following equation:

$$\frac{\partial c}{\partial \tilde{t}} = \frac{\partial^2 c}{\partial \xi^2} - kc \tag{14}$$

where k is obtained through the linearization of the equation of the substrate consumption rate (e.g. starting from the Monod law). Discretizing the spatial coordinate ξ in m finite zones, Equation (14) is transformed in the following system of differential equations:

$$\frac{\mathrm{d}c_j}{\mathrm{d}t} = m^2\left(c_{j-1} - 2c_j + c_{j+1}\right) - kc_j, \quad j = 1,2,\ldots,m \tag{15}$$

Considering the limit conditions (8), it results:

$$c_1 = c_0 = 0, \quad c_{m+1} = c^b \tag{16}$$

and the model of the concentration in biofilm becomes:

$$\begin{aligned}
\frac{\mathrm{d}c_1}{\mathrm{d}t} &= -\left(m^2 + k\right)c_1 + m^2 c_2 \\
\frac{\mathrm{d}c_2}{\mathrm{d}t} &= -\left(2m^2 + k\right)c_1 + m^2 c_2 + m^2 c_3 \\
&\cdots\cdots\cdots\cdots\cdots\cdots \\
\frac{\mathrm{d}c_m}{\mathrm{d}t} &= -\left(2m^2 + k\right)c_{m-1} + m^2 c_m + m^2 c^b
\end{aligned} \tag{17}$$

In what follows, it was adopted $m = 12$. For the discretization of spatial coordinate z, three finite elements ($N = 3$) were considered. Equation (13) can be written for each finite elements, in which the liquid concentrations are c_1^b, c_2^b, c_3^b. The terms $\left[\frac{\partial c}{\partial \xi}\right]_{\xi=1}$ come from the distinct discretized models of the biofilm, corresponding to the three finite elements. Denoting with k the current finite element ($k = 1, 2, 3$), the factor concerned may be written as follows:

$$\left[\frac{\partial c}{\partial \xi}\right]_{\xi=1} \cong \frac{c_m^{(k)} - c_{m-1}^{(k)}}{(1/m)} \tag{18}$$

A pulse was applied to the input of the simulated biofilter and the response obtained is shown in **Figure 7**. In this figure, the curves plotted for $k = 1$, $k = 2$ and $k = 3$ represent the responses obtained to the outputs of finite elements 1, 2 and 3, respectively ($k = 3$ corresponds to the biofilter output).

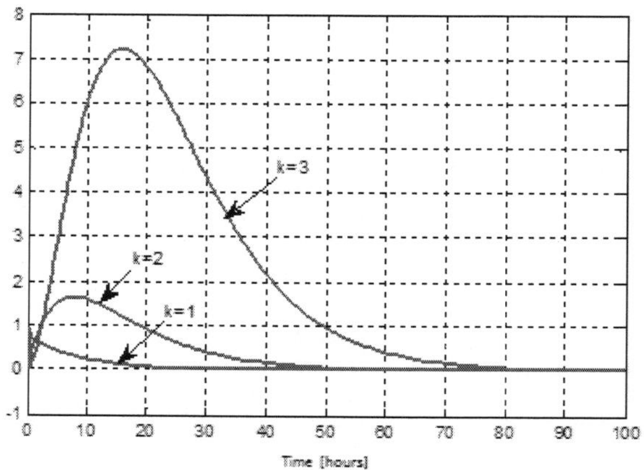

Figure 7. Pulse responses of the elements k from the biofilter structure (the case of linear model).

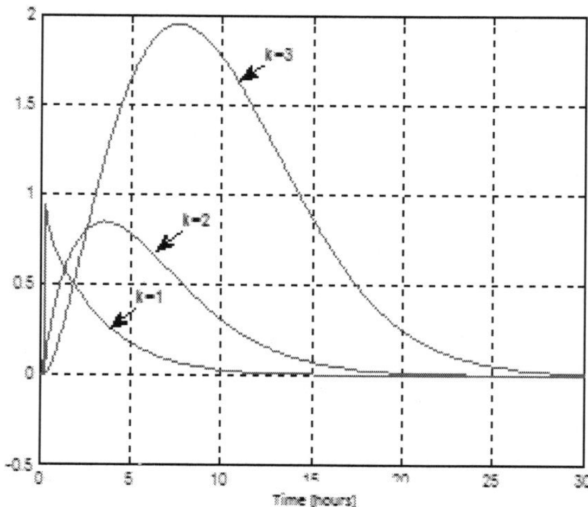

Figure 8. Pulse response of the elements k from the biofilter structure (the case of non-linear model).

It is now considered the non-linear case of the concentration model in biofilm in which, in Equation (7), the consumption rate of the component *c*, *r*(*c*), has a given parameterization of the Monod type, such that the concentration model in biofilm becomes non-linear. **Figure 8** shows the pulse responses obtained to the outputs of the finite elements 1, 2 and 3, respectively.

Remark: The numerical methods used before transform partial differential equations in ordinary differential equations. The main advantage of these methods is that they can also be used in the case of non-linear systems, allowing the use of any type of analytical expression for the substrate consumption rate. The drawback of numerical methods is that they do not allow the obtaining of traditional models of transfer function type, Bode characteristics etc., used in usual control structures. Instead, by their means, internal model-based control (IMC) structures can be implemented.

In the software packages for modeling and numerical simulation of the biofilters, such as AQUASIM [8], the network method is used. It involves the simultaneous discretization of temporal and spatial coordinates. The model of trickling biofilter was implemented and its parameters were identified using the existent functions in AQUASIM. For simulations, a structure of trickling biofilter similar to the one shown in **Figure 6**, with $N = 5$ zones, was considered. The model implementation started from the fact that in the case of RAS, the main component of the wastewater reaching the trickling biofilter is ammonia, the organic substrate being negligible. Four processes that occur in the nitrifying biofilter of trickling type were considered. **Table 3** presents the reaction kinetics and stoichiometric coefficients of these processes, they being in accordance with the activated sludge model (ASM) [9].

Variable / Process	Dissolved oxygen, S_{O_2}	Ammonia S_{NH_4}	Autotrophic biomass, X_A	Inert biomass X_I	Reaction kinetics
Autotrophic growth	$1-{}^{4.75}/_{Y_A}$	${}^{1}/_{Y_A}$	1	0	$\mu_A \frac{S_{O_2}}{K_{A,O_2}+S_{O_2}} \cdot \frac{S_{NH_4}}{K_{A,NH_4}+S_{NH_4}} X_A$
Autotrophic inactivation	0	0	–1	1	$k_A X_A$
Autotrophic maintenance	–1	0	–1	0	$b_A \frac{S_{O_2}}{K_{A,O_2}+S_{O_2}} X_A$
Aeration	1	0	0	0	$K_L\, a\left(S_{O_{2,sat}} - S_{O_2}\right)$

Table 3. Reaction kinetics and stoichiometric coefficients of the model implemented in AQUASIM.

The model of trickling biofilter was simulated considering the parameters in accordance with those of Activated Sludge Model No. 1 (ASM1). The obtained results are shown in **Figure 9** **[10]**, where it can be noticed that the biofilter reaches the steady-state regime. This simulation was necessary because all data are supplied by aquaculture pilot plant when the trickling biofilter operates in the steady-state regime. The simulation considered that the biofilter has an initial thickness of 1 micron, corresponding to the thickness of a particle. Practically, it presents the result of the biofilm formation, observing that the system goes into the steady-state regime after about 120 days. The evolution of the main components, ammonia and dissolved oxygen at the output of the three zones of the biofilter (Zones 1, 3 and 5) are shown in Figure 9a and 9b, respectively. Figure 9c shows the graphical representation of ammonia concentration with respect to the biofilter thickness at the end of the simulation time. It can be noticed that in the points of interaction with the liquid volume, the ammonia concentration in biofilm is equal to the one from the liquid volume, and it decreases toward the inside of the biofilm. Figure 9d shows the evolution of the biofilm thickness in the three zones mentioned before. It can be observed that the evolution of the biofilm thickness inside the biofilter is determined by the decrease of ammonia concentration from water along the height of the biofilter.

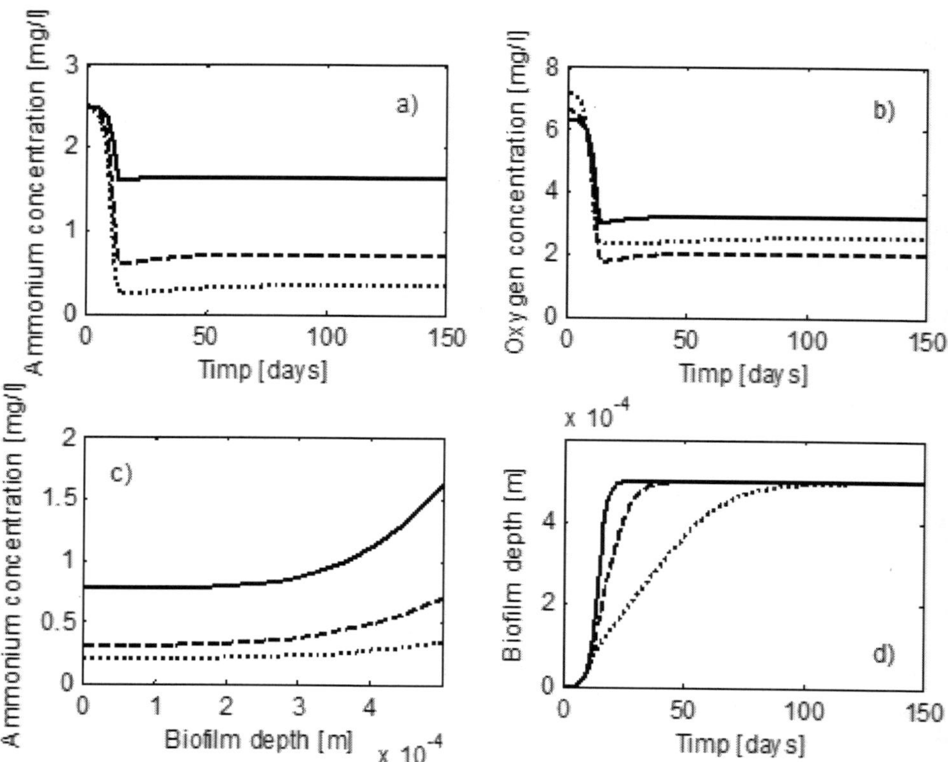

Figure 9. Simulation of the analytical model of trickling biofilter (Zone 1—solid line, Zone 3—dotted line and Zone 5—dashed line): (a) ammonia concentration in liquid volume; (b) dissolved oxygen concentration in liquid volume; (c) profile of ammonia concentration along the biofilm thickness; and (d) evolution of the biofilm thickness in trickling biofilter [10].

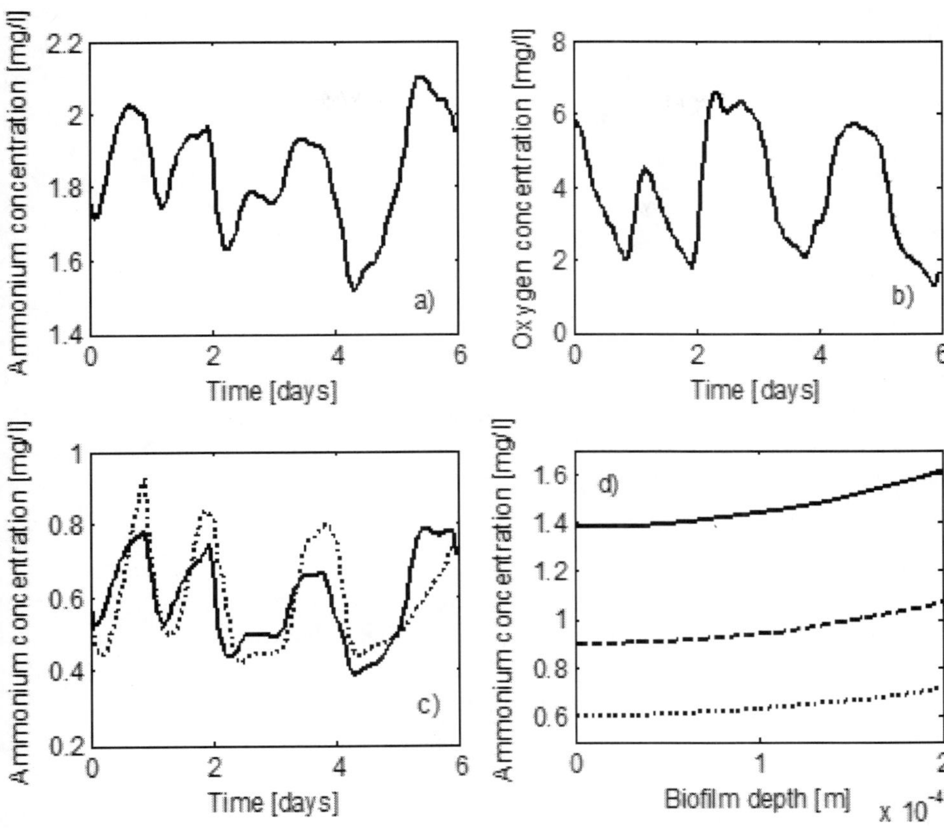

Figure 10. Simulation results of the identified model of trickling biofilter: (a) ammonia concentration in the influent; (b) dissolved oxygen concentration in the influent; (c) ammonia concentration in the effluent (experimental data—dashed line, evolution of the identified model—solid line); and (d) ammonia concentration along the biofilm thickness (Zone 1—solid line, Zone 3—dotted line and Zone 5—dashed line) [10].

The major advantage of this model implemented in AQUASIM is that it provides a detailed description of the phenomenology that takes place in the trickling biofilter. Thus, the model was also used as emulator to generate data from biofilter in other modeling studies.

A solution to obtain a simpler mathematical model is the modeling of trickling biofilter using an adaptive filter. Although the trickling biofilter is a non-linear system with distributed parameters, for control goals is sufficient to know its linearized mathematical model around the current operating point. Obviously, if the operating point of the biofilter changes, it is necessary to determine the updated linear model. In these conditions, the trickling biofilter can be treated as a variant dynamic system with distributed parameters [7]. A powerful tool to identify these systems is the adaptive filter.

Let us consider $h_a(t)$ and $h_o(t)$ the pulse responses of the biofilter on the channels $NH_{4,in}(t) \rightarrow NH_{4,out}(t)$ and $Q_{in}(t) \rightarrow NH_{4,out}(t)$, respectively. Based on the samples of the pulse responses $h_a[k]$ and $h_o[k]$, where k is the discrete time, the vectors of pulse responses $\boldsymbol{h}_a[\boldsymbol{k}]$ and $\boldsymbol{h}_o[\boldsymbol{k}]$ are formed. By noting $\mathbf{h}[k]=\left[\mathbf{h}_a^T[k] \quad \mathbf{h}_o^T[k]\right]^T$ and the samples of ammonia concentration and the inflow with $\mathbf{x}[k]=\left[\mathbf{x}_{NH_{4,in}}^T[k] \quad \mathbf{x}_{Q_{in}}^T[k]\right]^T$, the process model can be written as follows:

$$y[k]=\mathbf{h}^T[k]\mathbf{x}[k] \tag{19}$$

where $y[k] = NH_{4,out}[k]$.

The adjustment of the parameter vector, $\mathbf{h}[k]$, is done with the well-known recursive least square (RLS) algorithm [11]. On the basis of pulse responses $\boldsymbol{h}_a[\boldsymbol{k}]$ and $\boldsymbol{h}_o[\boldsymbol{k}]$, determined with RLS algorithm, the frequency characteristics of the process can be obtained. They represent the starting point of the methodologies of interactive frequency design of the control algorithms of trickling biofilter.

In the case of trickling biofilter, there are three variables that can modify the operating point: the feed flow rate of trickling biofilter (which actually is the recirculating flow), Q_{in}; ammonia concentration from the influent of trickling biofilter (which actually is the ammonia concentration in aquaculture tanks), $NH_{4,in}$; and dissolved oxygen concentration in the water treated in trickling biofilter (determined by the dissolved oxygen concentration in aquaculture tanks and the aerating processes from trickling biofilter).

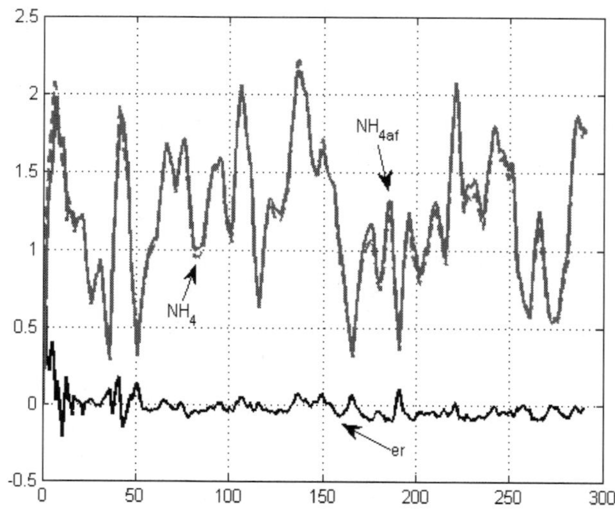

Figure 11. Validation of the identified model using adaptive filters (model output—blue line, output of the emulated process—red line and identification error—black line).

To highlight the modification of the properties of the adaptive dynamic model when the operating regime of biofilter changes, two extreme operating regimes were considered:

- High flow: $Q_{in} = 4\ m^3$ and $NH_{4,in} = 2$ mg N/L;
- Low flow: $Q_{in} = 2\ m^3$ and $NH_{4,in} = 4$ mg N/L.

It can be noticed that in aquaculture plant, in the two operating regimes, the same amount of nitrogen can be found: 8 g. It can be also noticed that in the mentioned situation, a constant value of dissolved oxygen concentration was considered: $DO_{in} = 4$ mg O_2/l. The software implementation in AQUASIM of the analytical model determined by identification was used as process emulator, obtaining the process output in the three operating regimes. **Figure 11** shows an example of identification using adaptive filters. In **Figure 11**, a very good match can be noticed between the output of the identified model and the one of the emulated process.

Figure 12 shows the Nyquist frequency characteristics of the channel $Q_{in}(t) \rightarrow NH_{4,out}(t)$. Analyzing these characteristics, it can be seen that the water flow which supplies the trickling biofilter has a great influence on the process dynamics. The change of the flow leads to the change of the gain and time constants of the transfer function identified on this channel.

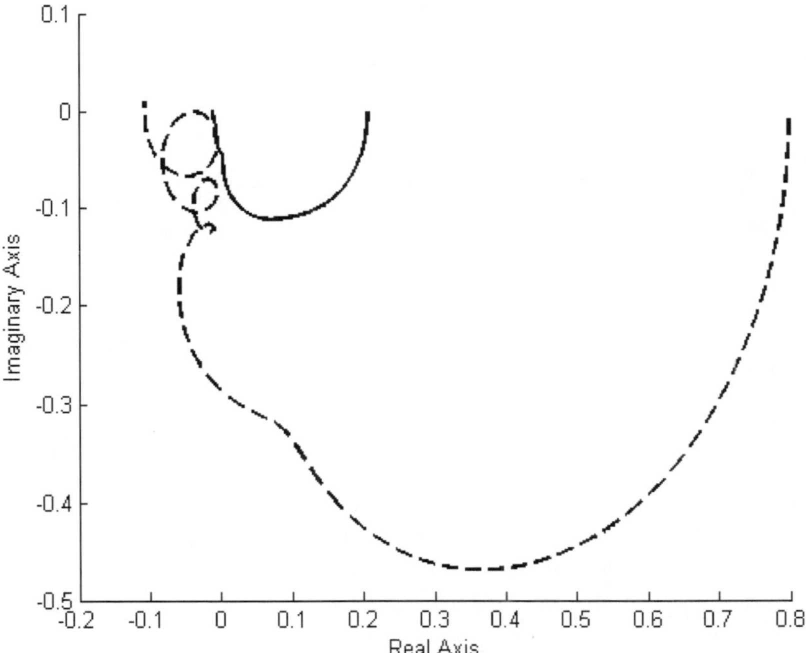

Figure 12. Nyquist frequency characteristic of the channel $Q_{in}(t) \rightarrow NH_{4,out}(t)$ (regime high flow—solid line and regime low flow—dashed line).

Figure 13 shows the Nyquist frequency characteristics of the channel $NH_{4,in}(t) \rightarrow NH_{4,out}(t)$. It is a disturbing channel, the ammonia concentration at the input of the trickling biofilter being determined by metabolic processes that take place in aquaculture tanks. From the analysis of the figures previously presented, it can be noticed that, despite a significant influence of this channel on the output in the two operating regimes, it has similar dynamic properties at low and medium frequencies.

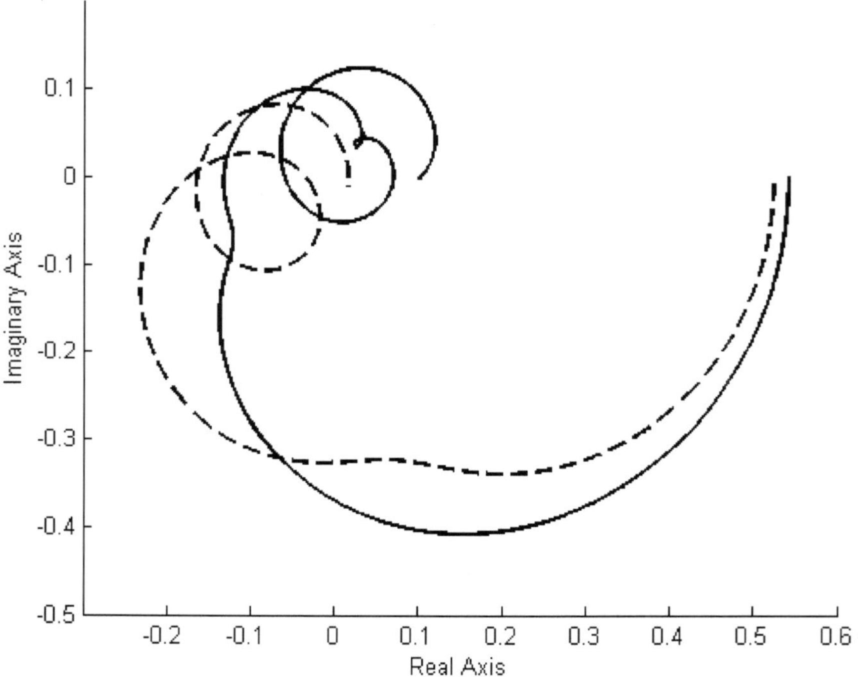

Figure 13. Nyquist frequency characteristic of the channel $NH_{4,in}(t) \rightarrow NH_{4,out}(t)$ (regime high flow—solid line and regime low flow—dashed line).

Finally, an analysis of the dynamic properties of the channel $DO_{in}(t) \rightarrow NH_{4,out}(t)$ was performed, and the obtained results showed that there is not a significant dynamics of this channel in the frequency domain of interest.

This analysis highlighted that the main control variable existent in the case of trickling biofilters is the recirculated flow. The analysis also showed that the dynamic properties of the control channel $Q_{in}(t) \rightarrow NH_{4,out}(t)$ vary greatly with respect to the operating point, from the two points of view: gain and time constants [7]. Thus, it results the necessity to use robust control techniques by the approximation of this channel with variable parameter linear models. The control of the recirculated flow can be performed directly if the aquaculture plant is equipped with variable flow recirculating pumps or indirectly through the control of the water level in aquaculture tanks.

In the case of the control channel $DO_{in}(t) \rightarrow NH_{4,out}(t)$, the lack of significant dynamics was highlighted. At the same time, in practical investigations on the pilot plant, it can be noticed that the use of the control to aerate the trickling biofilter does not lead to satisfactory results [7]. In these conditions, the indirect control of dissolved oxygen concentration in the water inside the trickling biofilter can be done through the direct control of dissolved oxygen concentration in the water from aquaculture tanks.

In order to design a control system of trickling biofilter must also take into account the disturbing channel $NH_{4,in}(t) \rightarrow NH_{4,out}(t)$. This channel is influenced by the food introduced into aquaculture plant and the metabolic processes of the fish population [7]. These processes represent the determining factors in establishing the operating mode of a recirculating aquaculture plant. Depending on ammonia concentration in aquaculture tanks, the recirculating flow within the plant is set. Thus, it seeks to establish inside the plant an ammonia concentration of the water of maximum 1 mg N/L. For the disturbing channel $NH_{4,in}(t) \rightarrow NH_{4,out}(t)$, techniques of feed-forward type or robust control can be used, if this disturbance is not measurable.

4. Experimental results regarding RAS dynamics and control

To emphasize the dynamic properties of RAS and the control solutions, an experiment in which a species having an intensive metabolism (*Cyprinus carpio*) was carried out. Thereby, a high ammonia concentration was obtained in this experiment. **Table 4** presents the biomass distribution in the tanks of RAS [12].

Tank number	Mass (kg)	Number of individuals	Average mass/individual (g)
1	$C_1 = 13{,}616$	665	20.47
2	$C_2 = 13{,}614$	557	24.44
3	$C_3 = 13{,}855$	614	22.61
4	$C_4 = 13{,}710$	591	23.19

Table 4. The populating mode of aquaculture tanks [12].

The fodder Optiline 1 P of 2 mm having 44% content of protein [13] was used for feeding the fish biomass.

The data were collected from the process during two experiments: the first experiment used a continuous distribution of the fodder, and the second a discontinuous one giving three ratios per day (at 9:30, 14:00 and 18:30). The first experiment was lengthy, and it used a sample period of 10 minute; the second was of shorter duration, with 1 minute sampling period.

The analysis of the process data in order to monitor RAS highlights that all physical variables are affected by an important high-frequency noise which imposes the use of an efficient filtering system. In the developed monitoring system, the filtering subsystem is composed of two units in series: a non-linear filter for the removal of the important short-duration variations and a classic linear filter of second order for the ordinary high-frequency noise. Figure 14a shows the effect of the filtering system to the acquisition of a signal given by ammonia sensor from the biological filter.

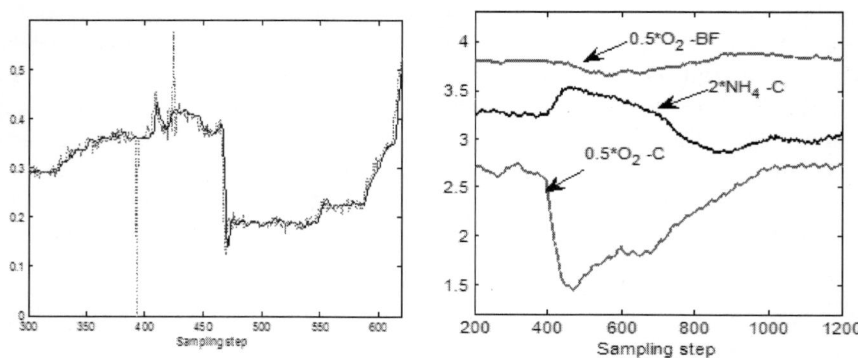

Figure 14. (a) Sample of ammonia concentration at the biological filter output, NH_4-BF (mg/L): unfiltered (dot) and filtered signal (solid); (b) evolution of ammonia concentration at the biofilter input, NH_4-C (mg/L) (red) and of oxygen concentrations at the biofilter input and output, respectively, O_2 and O_2-BF (mg/L) (red and blue, respectively).

Together with high-frequency disturbances, the collected signals may be affected by a slow drift due to the deposition of biofilm on the sensitive surfaces of the sensor from the liquid medium. Therefore, it was necessary to apply a careful maintenance to reduce these errors.

The main disturbance that affects the acquired variables from RAS is the one resulted from the fish feeding. This has two components: the first composed of dejections and metabolism products, which constitute the main component, and the second – the organic substances resulted from the decomposition of the unconsumed fodder. Figure 14b shows the variations of the oxygen concentrations inside the aquaculture tanks (O_2), the biological filter (O_2-BF) and the ammonia concentration collected at the biofilter input (NH_4-C) considering a 1 minute sample period, when the fodder is given in batch mode. After about 3 hours from the feeding, the ammonia concentration increased fast, and then, after 10–11 hours, the concentration returned to the initial concentration, as a result of the action of the biological filter and denitrificator from the recirculated water circuit [12]. Figure 14b shows that, at the same time with the increase of ammonia concentration, a pronounced decrease of the oxygen concentration in the aquaculture tanks takes place. The effect on the oxygen concentration at the biofilter output is much lower.

The internal pseudo-periodical disturbances have an important weight in the RAS dynamics. They are generated by the washing processes of mechanical, sand and carbon filters [4, 12]. The wash of the mechanical filter is accompanied by a loss of water removed from the system together with the slime, which causes a sudden decrease of the water level in aquaculture tanks.

At the same time, the wash of sand and active carbon filters is achieved through their bypass. When the filters are recoupled in the circuit, a sudden decrease of the water level in aquaculture tanks occurs. In both cases, the systems that provide the imposed water level in the tanks perform the compensation of water losses through an intake from the water network. The internal disturbances produced by the cyclic operating of mechanical, sand and active carbon filters generate a complex dynamics of RAS when it operates in permanent regime. Analyzing this dynamic regime offers useful information for the system control. Thus, **Figure 15** shows the evolutions of ammonia and oxygen concentrations at the biofilter's input and output (NH_4-C and NH_4-BF, O_2, and O_2-BF, respectively). These variations show the biofilter efficiency through the significant difference between ammonia and oxygen concentrations at the biofilter's input and output. Obviously, the two types of physical variables have evolutions, mostly, in anti-phase.

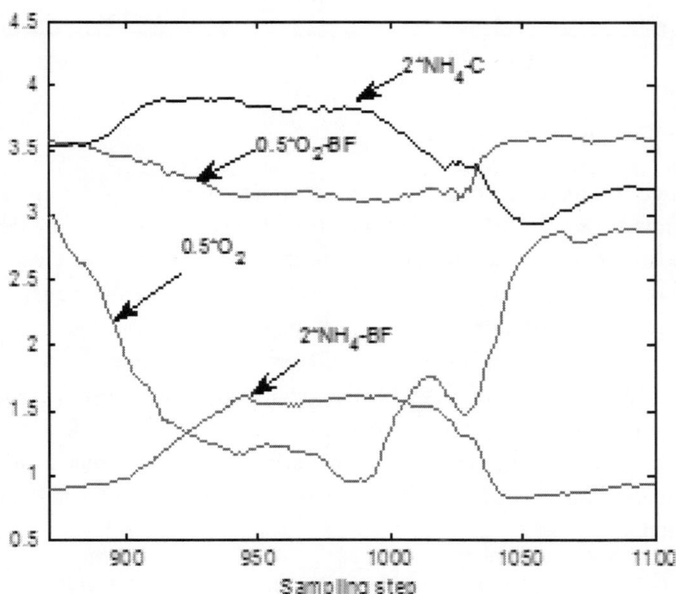

Figure 15. Evolutions of ammonia concentrations at biofilter input and output, NH_4-C and NH_4-BF (mg/L) (black and green, respectively) and of oxygen concentrations at biofilter input and output, O_2 and O_2-BF (mg/L) (red and blue, respectively).

As shown in Section 2, the control of RAS is structured into two hierarchical levels. The first level performs data acquisition, their processing according to the necessities of monitoring and control functions (in this case, the operation of disturbance and high-frequency noise removal, which have an important weight, is essential), and the control loops. These loops are referring to the water level and oxygen concentration in aquaculture tanks and to pH control in the collecting tank located to the output of mechanical filter. **Figure 16** shows the response of pH control system when the operating regime switches acid/base.

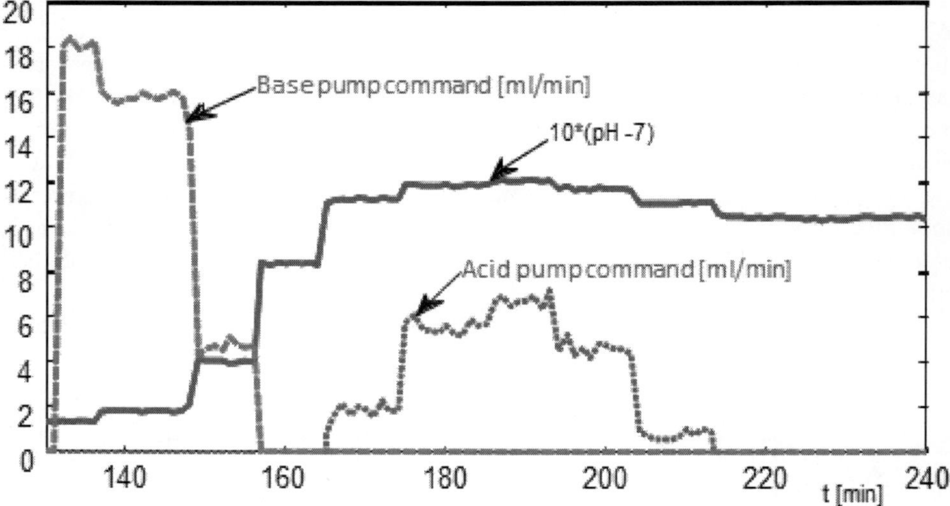

Figure 16. The response of pH control system.

For economic reasons, the water circulation in RAS is achieved by two pumps with constant flow. Both pumps are controlled in on–off regime by the controllers that provide a constant level in the two collecting tanks located after the mechanical filter and after trickling filter. This solution does not allow the direct control of the recirculated flow in the aquaculture system. The flow adjustment can be done through the average level imposed in the aquaculture tanks. Figure 17a shows the correlation between the evolution of the average level, L, and the total inflow in aquaculture tanks, IF_f. In this graph, the signal IF_f is obtained using a moving average filter, to whose input the sum of the inflows in the aquaculture tanks is applied. For RAS operating necessities, a nomogram determined experimentally from which the water level set point in aquaculture tanks is deduced, aiming to obtain a desired adjustment of the recirculated flow is used.

To control the nitrification process through the trickling filter, some solutions were investigated, the first being the use of the recirculating flow as control variable. Generally speaking, the increase of the recirculating flow leads to the decrease of ammonia concentration in the aquaculture tanks. However, the domain of the recirculating rate is limited both in terms of technologically and also due to the cost of the consumed electrical energy. The control of the nitrification process is practical compromised because of strong variations of the physical variables of the system that are produced by the washing processes of mechanical, sand and active carbon filters. Figure 17b shows a fragment from a record of the recirculating flow affected by two consecutive washes of a filter, together with the corresponding variation in ammonia concentration at trickling filter output. It is obvious that internal disturbances from RAS make it difficult to discern the effects of the control applied to the recirculating flow by the variations induced through these internal disturbances.

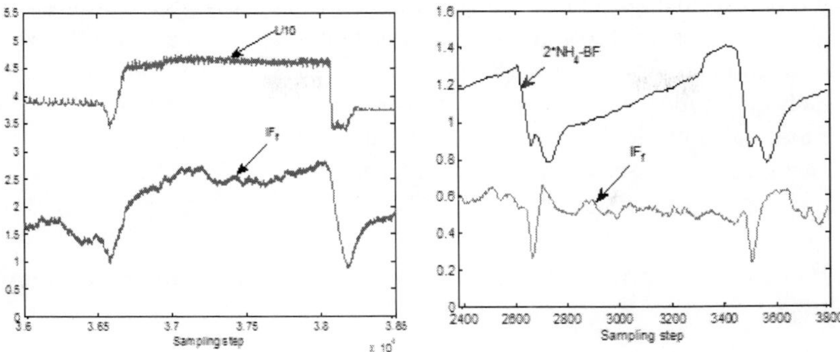

Figure 17. (a) Evolution of the level in the tank located after mechanical filter, L (cm), and of total inflow of the aquaculture tanks (filtered), IF_f (m^3/h); (b) evolution of recirculating flow, IF_f (m^3/h) affected by two consecutive washes of filters, together with the corresponding variation of ammonia concentration in biofilter NH_4-BF (mg/L).

The opportunity to increase the biofilter efficiency was also analyzed by an aerating process in countercurrent to the flow of the processed water. It was found that the aerating control of the biofilter has practically a negligible effect so that such a solution is not appropriate. Because of spaces between balls inside the trickling biofilter, a sufficient natural aeration is produced, which excludes a supplementary aerating system [12]. An indirect control solution of the nitrification process is suggested by the connection between oxygen concentration in the aquaculture plant and ammonia concentration at the biofilter output, NH_4-BF (**Figure 18**). If necessary, the reducing of ammonia concentration at the biofilter output can be performed by a control aiming to intensify the aeration of aquaculture tanks.

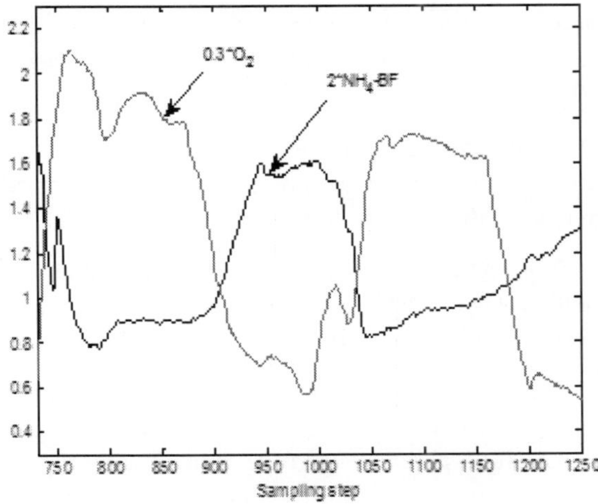

Figure 18. Evolution of ammonia concentration at biofilter output, NH_4-BF (mg/L) (black) and of oxygen concentration in aquaculture tanks, O_2 (mg/L) (red).

As a conclusion, due to the fact that the trickling filter does not have proper means of control, the nitrification process can be controlled only through the recirculating flow or, if necessary, through aeration intensification of aquaculture tanks. The high level of system's disturbances, especially those produced by the operation of mechanical, sand and carbon filters, requires the use of the qualitative description of the essential variables of RAS, which is presented in this section, to create a rule-based system for RAS control.

5. Expert system for monitoring and control of the recirculating aquaculture process

For monitoring and overall control of the aquaculture recirculating process, an expert system that uses the human expertise in the aquaculture field has been implemented. It contains over 200 rules and performs the following functions: monitoring of data acquisition process, monitoring of control loops operation, monitoring the chemistry of the water recirculating system, the establishing of nutrition strategy, and monitoring of technological performance. The expert system relies on production rules and it consists of the following modules: the database, the base of facts, the knowledge base, the inference engine and the HMI.

The database can be viewed as an auxiliary memory that contains the following data:

- online measured data. These data are provided by the process sensors considering a sample period equal to 1 minute. The following measured variables can be mentioned: temperature current value, oxygen concentration current value, water level current value in the aquaculture tanks, ammonium concentration current value etc.
- data initialized by operator (the absolute minimum or maximum value of the oxygen concentration in the aquaculture tanks, the maximum admissible value of the ammonium concentration in the biological filter etc.)
- data provided by the operators or calculated during the process. The operator should introduce every 2 weeks the total number of individuals from the tank B_i, $i = 1, 4$, the weight of each individual etc. On this basis, the following variables are calculated: the total fish biomass at the current step, the SGR, the food conversion factor etc.

The base of facts: The facts are defined as simple sentences as follows: water temperature = 15°C, O_2 concentration in the tank $i = 3$, NH_4 concentration in the biological filter = 0.15 (mg/L), NO_3 concentration = 50 mg/L etc.

The knowledge base: It contains a production rules network of ***IF****premise****THEN****conclusion/Prio/Comment* type. The following sources have been used for the knowledge acquisition: interviewing the specialist, making experiments and literature. The premise may contain logical operators of AND, OR, NOT type. The conclusion may signal a certain state of the process and may suggest an appropriate action. Furthermore, some examples of rules are presented:

- Monitoring rules of data acquisition

IF {O2cM–O2cBi > 2} **THEN** {The O_2 sensor from the tank ***i*** is faulty or the tank aeration is damaged}/Prio N/**Comment**: Check and eventually calibrate the sensor. The blower can be disconnected through thermal protection or the blower flow be insufficient,

where O2cM is the maximum value of the oxygen concentration, O2cBi is the current value of the oxygen concentration and Prio N is the normal priority.

IF {NH4cC–NH4mFB < 0.5} **AND** { NH4cC–NH4mFB > 0} **THEN** {Faulty evolution of the nitrification process}/**Prio H**/**Comment**: Check the biofilter operation (if the liquid flux is uniformly distributed on the biofilter section),

where NH4cC is the current value of the ammonium concentration, NH4mFB is the ammonium concentration measured in the last hour inside the biological filter and Prio H is the high priority.

- Monitoring rules of the chemistry of recirculating water system

IF {pHc ≤ 6,5} **THEN** {The pH is less than the admissible limit}/**Prio N**/**Comment**: Check if there is alkaline agent in the control loop AND increase the water recirculating flow OR increase the water refresh flow,

where pHc is the current value of pH.

- Rules for establishing the feeding strategy

IF {GMI*i* ∈ (10–50 g)} **AND** {TcB*i* ∈ (18, 20°C)} **THEN** {The food rate in the tank *i* is (3%/MC*i*)}/ Prio **N**/**Comment**: Set the food rate for the fish of this age category,

where GMI***i*** is the average weight of the individual in the tank *i*, TcB*i* is the current value of the temperature in the tank *i* and MC*i* is the body mass of an individual in the tank *i*.

- Monitoring rules of the technological performance

IF {GMI*i* ∈ (50–200 g)} **AND** {SGR <3%} **THEN** {The fish biomass in the tank ***i*** does not develop normally}/**Prio N**/**Comment**: Check the technological conditions AND/OR the food rate should be adapted,

where GMI*i* is the average weight of the individual in the tank *i*, SGR is the specific growth rate, $SGR_k = 100\ (\ln B_k - \ln B_{k-1})/t$, where *t* is the time between the last two weighing.

The inference engine: In the present control application, a forward-chaining strategy, specific to the expert systems based on production rules, was used. The reasoning is of deductive type, from the facts to the goal.

HMI: The process is operated by a friendly graphical interface that communicates with the expert system. The main screen of the interface (**Figure 19**) contains a synoptic scheme of the process where it can be seen the global state of the process [14]. HMI gives to the operator the possibility to visualize online the main variables of the process, to control the process in manual or automatic regime, to plot and to store the values of the process variables for later processing.

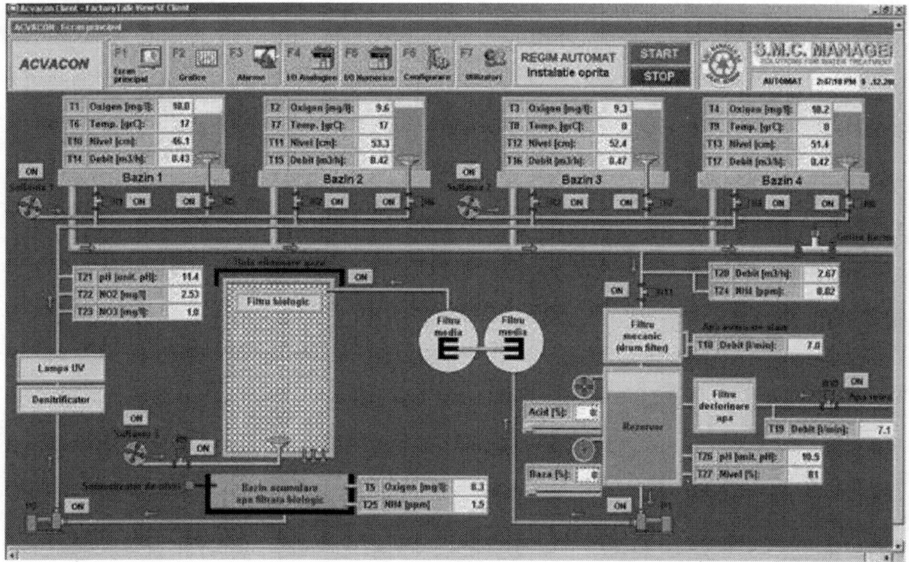

Figure 19. The main screen of the interface [14].

6. Conclusions

The performance of the biological filtering of the recirculated water in RAS has a crucial importance because it provides a proper hygienic and sanitary state of culture biomass. As a last resort, the ratio between the fish production and the culture space provided by RAS depends on this performance. The analysis of the pilot plant of RAS containing a trickling biofilter and a chemical denitrificator, made in "Dunarea de Jos" University of Galati, mainly targeted aspects of modeling, monitoring and control of the pilot station. It was confirmed experimentally that the biofilter aeration in countercurrent with respect to the flow of the processed water has a practical negligible effect so that the trickling biofilter does not offer control means of the nitrification process. In these conditions, the main possibility to control the nitrification process is the control of recirculated flow. The practical expertise from the operation of the pilot plant showed that the use of the recirculating pumps having a reduced cost, with constant flow and on–off control, is not an adequate solution in terms of energy consumption and mostly from the control necessities point of view. An alternative to this solution is the use of variable flow pumps, driven by frequency controlled asynchronous motors. An indirect control solution of the nitrification process that can be applied to reduce ammonia concentration at the biofilter output consists in the aeration intensification in aquaculture tanks.

The main difficulty of RAS control is generated by the disturbances that strongly affect all the system variables. These disturbances are produced by the washing processes of mechanical, sand and active carbon filters of RAS. Their presence makes difficult to discern the effects of control applied to the recirculating flow by the variations induced through internal disturbances. A possible solution, validated experimentally, consists in reducing these disturbances through the removal of sand and active carbon filters. It has been shown that their presence is not essential. In this case, the frequency of the internal disturbances, induced only by mechanical filter, is significantly reduced, having positive effects on the conditions in which the process control is carried out. The monitoring and control system of RAS provides, at the first hierarchical level, the local control loops of the process variables: water level, O_2 concentrations and pH. At the superior level, the process monitoring, its operating, and the control of nitrification process are achieved. Because of the high level of the system's disturbances, it was considered that the best solution for achieving these functions is to use an expert system.

An integrated modeling of RAS was performed, taking into account the phenomena that take place both in the biological subsystem of the fish population and at the level of the microbiological subsystem of the water quality. Thus, a distributed parameter system based on partial differential equations of the biofilm formation and substrate consumption and on reaction kinetics of ASM1 type was obtained. The model has been identified based on the experimental data taken from aquaculture pilot plant located in "Dunarea de Jos" University of Galati. This model was used as process emulator for a complete analysis of trickling biofilter. It allowed testing its behavior through numerical simulation in different situations, some of them being very difficult to obtain practically, because it could affect the fish biomass or even the microorganism population. This also allowed the treating of the biofilm through an adaptive filter, allowing the sensibility analysis of frequency models for each I–O channel.

References

[1] Timmons M.B., Ebeling J.M. Recirculating aquaculture. 3rd Edition. Ithaca Publishing Company; Ithaca, NY 2013.

[2] Ebeling J.M., Timmons M.B. Recirculating Aquaculture Systems. In: Tidwell J.H., editor. Aquaculture Production Systems. Wiley-Blackwell; Oxford; 2012.

[3] Timmons M.B., Ebeling J.M. Recirculating Aquaculture. Cayuga Aqua Ventures; Ithaca, NY 2007.

[4] Barbu M., Ionescu T., Ifrim G., Caraman S., Cristea V., Ceanga E. Results regarding the water quality control in recirculating aquaculture systems. Journal of Environmental Protection and Ecology. 2012; 13 (1): 39–47.

[5] Wik T., Linden B., Wramner P. Integrated dynamic aquaculture and wastewater treatment modelling for recirculating aquaculture systems. Aquaculture. 2009; 287 (3–4): 361–370.

[6] Wanner O., Eberl HJ., Morgenroth E., Noguera D., Picioreanu C., Rittmann BE., Van Loosdrecht M.C.M. Mathematical Modeling of Biofilms. IWA Publishing, London; 2006.

[7] Barbu M., Mînzu V., Carp D., Ceanga E. Identification and sensitivity analysis of a trickling biofilter viewed as a distributed parameters system. In: 15th International Conference on System Theory, Control and Computing, ICSTCC, Sinaia, Romania; 2011.

[8] Reichert P. Design techniques of a computer program for the identification of processes and the simulation of water quality in aquatic systems. Environmental Software. 1995; 10 (3): 199–210.

[9] Henze M., Gujer W., Mino T., van Loosdrecht M.C.M. Activated Sludge Models ASM1, ASM2, ASM2d and ASM3. IWA Publishing, London; 2000.

[10] Barbu M., Picioreanu C. Model of trickling biofilter from an intensive recirculating aquaculture system. In: IWA Biofilm Conference: Processes in Biofilms, Shanghai, China; 2011.

[11] Haykin S. Adaptive Filter Theory. Prantice Hall; Englewood Cliffs, NJ 1995.

[12] Caraman S., Barbu M., Ionescu T., Ifrim G., Cristea V., Ceanga E. The analysis of the dynamic properties of the wastewater treatment process in a recirculating aquaculture system. Romanian Biotechnological Letters. 2010; 15 (4): 5457–5466.

[13] Barbu M. Experimental results regarding the operating regimes of trickling filters in recirculating aquaculture systems. Fresenius Environmental Bulletin. 2012; 21 (11c): 3500–3506.

[14] Barbu M., Caraman S., Vlad C., Nicolau T., Ceang E. Hierarchical control system for recirculating aquaculture processes. In: 16th International Conference on System Theory, Control and Computing, ICSTCC 2012, Sinaia, Romania; 2012.

Index

Population movement 90